Hesiod's Anvil

falling and spinning
through
heaven and earth

© *2007 by*
The Mathematical Association of America (Incorporated)
Library of Congress Catalog Card Number 2006933945
ISBN 978-0-88385-336-8
Printed in the United States of America
Current Printing (last digit):
10 9 8 7 6 5 4 3 2 1

The Dolciani Mathematical Expositions

NUMBER THIRTY

Hesiod's Anvil

falling and spinning through heaven and earth

Andrew J. Simoson
King College

Published and Distributed by
The Mathematical Association of America

The DOLCIANI MATHEMATICAL EXPOSITIONS series of the Mathematical Association of America was established through a generous gift to the Association from Mary P. Dolciani, Professor of Mathematics at Hunter College of the City University of New York. In making the gift, Professor Dolciani, herself an exceptionally talented and successful expositor of mathematics, had the purpose of furthering the ideal of excellence in mathematical exposition.

The Association, for its part, was delighted to accept the gracious gesture initiating the revolving fund for this series from one who has served the Association with distinction, both as a member of the Committee on Publications and as a member of the Board of Governors. It was with genuine pleasure that the Board chose to name the series in her honor.

The books in the series are selected for their lucid expository style and stimulating mathematical content. Typically, they contain an ample supply of exercises, many with accompanying solutions. They are intended to be sufficiently elementary for the undergraduate and even the mathematically inclined high-school student to understand and enjoy, but also to be interesting and sometimes challenging to the more advanced mathematician.

MAA Service Center
P.O. Box 91112
Washington, DC 20090-1112
1-800-331-1MAA   FAX: 1-301-206-9789

Contents

Dedicated to

Edward Walter Burke, Jr.

professor emeritus of physics
at King College,
who mentored me in the art
of telling stories in the classroom

Introduction

From catching balls on the playground, watching leaves swirl down in the fall, glimpsing shooting stars streak across the night, observing the planets move against the fixed stars, man has thought long and written often about falling and spinning motion, giving rise to models of the cosmos and formulas for motion in terms of space and time. This book recounts some of that heritage, blending literature and mechanics.

We quote Hesiod of ancient Greece and Plutarch of the Roman Empire on motion through heaven and earth. We consider gravity in Dante's *Inferno* and Edgar Allan Poe's *The Pit and the Pendulum*. We take stories of science fantasy by Jules Verne, Edgar Rice Burroughs, H. G. Wells, and Arthur C. Clarke, and use them to provide material on how objects fall and spin. These writers sometimes guessed rightly about motion, and sometimes wrongly. For example, Figure 1 is a pair of engravings from Verne's *From the Earth to the*

a. Before launch   b. In space

Figure 1. Free fall in 1870

Moon and *Round the Moon* showing his space capsule's interior before and after launch. Verne followed Johannes Kepler's conjecture that at some point between the earth and the moon, their gravity would exactly counterbalance one another, whereupon "the body will dangle immobile" [**44**, pp. 108–9, footnote 77]. That is, it was long thought that a feeling of weightlessness could only occur at very special points in space.

Anyone who has successfully taken a year-long calculus course should be able to understand most of this book, though even experts may find some surprises. We assume that a first year course may include an introduction to partial derivatives and basic multiple integration. Beyond first year calculus, a little knowledge of linear algebra and vector calculus is enough for almost everything in this book. When a mathematical tool or result beyond basic calculus is needed in a chapter, an explanation is given either in the text or in the appendix.

We assume that the reader is familiar with Newton's law of gravity that the attraction between two masses varies proportionately with their products and inversely proportionately with the square of the distance between them, and that an acceleration **a** on a mass m induces a force of $m\mathbf{a}$ on it. As much as possible we avoid using the terms of physics. For example, let m be the mass of an object moving along a number line, s be the distance of the object from the origin, v be its velocity, and a be its acceleration. Then $av = a\frac{ds}{dt} = v\frac{dv}{dt}$, which when integrated with respect to time, and when multiplied by m, gives

$$m \int a \, ds = \frac{1}{2}mv^2,$$

often called the work-kinetic energy equation since the left-hand side is a definition of work done in moving the object and the right-hand side is the definition of kinetic energy of the object. We use this equation later, but we will not need to talk about work done or kinetic energy. When we speak of momentum, inertia, centripetal acceleration, and so on, we will give some introduction.

This book is a series of preambles and chapters. The preambles are informal, giving personal anecdotes of falling and spinning. They may explain why or how the ensuing chapter was written, or give a metaphor of the kind of falling to be discussed. Heading the preambles are quotations, each containing at least one of the root words *fall, spin,* or *turn.* Although some of them may appear cryptic out of context, they form a long line through the book, demonstrating that people use the images of falling and spinning naturally. We *fall* in love; our stocks rise and *fall;* we *fall* in battle; our heads *spin*. Each of the preambles contains an endline doodle depicting a semblance of the story.

Each chapter takes an aspect of falling and analyzes it in the context of calculus, of rates of change, of accruing amounts, of highs and lows. Each chapter can be read independently of the others. The first and last chapters are bookends and give the book its title. At the ends of all but the first chapter are a few exercises, some straightforward, others open-ended. For some I have no answers. Some may make good student projects.

Versions of some chapters have appeared as my articles or notes over the years. Parts of Chapters II, VIII, X, and XIII appeared in *Mathematics Magazine*, [75], [76], [74], and [11]; parts of Chapter V and XII appeared in the *College Mathematics Journal*, [70] and [80]; Chapters VI and XI appeared in *Primus*, [78] and [77]; and Chapter VII appeared in the *American Mathematical Monthly*, [79].

The first chapter poses a few open-ended questions. How big is the universe? What kinds of terms are needful to describe the breadth of heaven and the depth of the earth? If you were a muse to the old Greek poets, how would you counsel them so that their descriptions of motion would ring true through the years? In the last chapter we give some advice to whisper in the poets' ears.

In between we explore aspects of falling and spinning. In Chapter II we consider models of the earth through the ages, and generate their gravitational acceleration functions in terms of distance from the earth's center. In Chapter IX we see how calculus and a little data can be used to determine the earth's structure.

We toss objects from the earth without resistance using Galileo's ballistic formulas in Chapter III and Newton's extension of them to heavenly motion in Chapter IV. In the process we look at a few early attempts to explain gravity and acceleration. As a case study we consider Verne's imaginative projectile shot to the moon one hundred years before NASA's Apollo program.

We explore pendulums in Chapter V and falling resistance in Chapter VI, analyzing stories by Poe and Wells where gravity is constant. Then in Chapters VII and VIII, we consider sliding and falling dynamics through the earth where gravity changes with distance. Chapter X discusses trochoids, curves that served as an early model for planetary motion. Chapter XII contrasts a space station's artificial gravity with the natural gravity of a small planet, using Arthur C. Clarke's spacecraft of *2001* and the Little Prince's asteroid planet B-612. In Chapter XI we consider Wells' gravity-blocking substance and show how it could be used to navigate through space. Chapter XIII considers the rotating beacon problem, a variation on a standard calculus exercise. Chapter XIV is a story problem interrelating the falling and spinning motions discussed in the other chapters.

By no means is this book comprehensive. There could be chapters on the gyroscope, the yo-yo, the n-body problem, the gravitational bending of light, and any of a host of motion-related phenomena, including Jonathan Swift's island in the sky, Laputa, of the Gulliver stories. Perhaps they may be another book.

The chapters can be read in any order. Inside the chapters, you can gloss over the derivations, jump to the prose and the conclusions, go back and recompute a trajectory, an acceleration, a velocity, a position, a time. You can use this book in the classroom or as a bedside item for thoughts on which to fall asleep. The idea of motion pervades the human experience, and has caused a great deal of controversy and excitement through the ages. The story, the people, the building up of ideas about falling and spinning is an intriguing strand in history.

Finally, I thank a number of people:

- Paul Scholten, a physicist, with whom I co-wrote my first motion paper [70] on the falling ladder paradox, the kernel of this book,

- Tom Barr, who commissioned me to write enrichment projects for a calculus text, thereby motivating me to translate some motion-related physics into the language of beginning calculus,

- Frank Farris, who as editor of the *Mathematics Magazine* chose two of my motion manuscripts as cover articles, and encouraged me to enlarge the work into a book,

- Jack Boyles, an artist, who allowed me to take his drawing course multiple times, and so is indirectly responsible for many of the sketches appearing herein,

- The *Penheads*, a writing group (three English professors, a psychologist, and me), who critiqued the first-draft prose of the text,

- Ray Bloomer and Dan Cross, physicists at King College, who welcomed my questions about physics over the years,

- Underwood Dudley, who meticulously edited the manuscript, yet encouraged freedom in style and content,

- Elaine Pedreira and Beverly Ruedi, MAA production editors, who transformed text and graphics into this book in real time,

- Connie, my wife, for letting me write long hours into the night.

July 2006
King College
Bristol, Tennessee

preamble i

I also say it is good to fall.

Walt Whitman (1819–1892), *Song of Myself*

left foot, right foot,

I was walking along a dusty trail.

left foot, right foot,

As I walk, I sometimes daydream, forgetting where I am, my feet taking care of themselves.

left foot, right foot,

The sand was the Kalahari. I had been teaching in Botswana for nearly a year. The last few months I had been working on a problem off and on. I knew I was close to a solution, if one existed.

left foot,

I felt tantalizingly close. But some problems can't be solved. My approach could be a dead-end. Maybe it was time to give up and try a fresh problem.

right foot,

A voice spoke. Audible? It seemed loud enough.

left

Its tone was patient, like the cadence of my step, and deliberate. My feet left impressions in the sand as when stepping through a sprinkling of snow.

foot,

"Look at your shoes," the voice said.

right

I looked—dusty shoes, scuffed and scratched by the steel-like thorns of the African shrubbery.

foot.

Then I saw my laces, their criss-crossing. In a moment, I knew that their pattern was the very one I needed to complete the argument to my problem. The solution had been there all the time at my feet.

I stopped and looked around. A goat met my gaze. No one was nearby. Whose voice had that been? The spirit of the gods of this place? The keeper of the book containing the best of mathematical arguments? My own sub-conscious mind?

Maybe this voice was like those of the Muses spoken of by Hesiod, a poet of old.

CHAPTER I

Hesiod's Muses

How distant is heaven? How deep is the earth?

How can we know such things? How can we even know enough to ask such questions?

When Isaac Newton was asked how he came to see how objects move in the cosmos, how he invented the calculus, he replied, "I stood on the shoulders of giants." His muse was the partial answers given by men such as Johannes Kepler, Galileo, and his mentor, Isaac Barrow.

But suppose there is no partial answer on which to build.

Imagine being an Adam or an Eve, the first humans, inventing language and logic, discovering apparent cause and effect, observing a meteorite fiery white against a black sky, and trying to explain to your mate in rational terms what has just occurred. It is natural to extrapolate from local awareness so as to form global perspectives. To Adam and Eve the world looked flat, so Eve might say to Adam,

> Earth is a great island in the midst of a sea. Below us the ground is deep. Above are lights ever so high. Like apples falling from the tree to the ground, bits of broken light fall to the earth.

Figure 1 is a rendition of this model as depicted in a current popular science fantasy series called Discworld, wherein all manner of strange folk tales come alive. This model was in vogue in the eastern Mediterranean basin in the days of Homer and Hesiod, except that the four elephants and the broad back of the sea turtle might be replaced by Atlas or merely darkness. Hesiod's greatest surviving work is the *Theogeny*, a poem detailing the lives of the gods: who bedded whom, who begat whom, where they made war, where they danced, and so on. The poem is a phone book of names for his listeners to invoke and fear. Hesiod drew upon a rich tradition of stories and legends, though he claimed that he received all of his supernatural

Figure 1. A disk world

information from the Nine Muses, "the ready-voiced daughters of Zeus," who whispered in his ear.

Nine? Why not ten muses? The number nine occurs often in the *Theogeny*. For example, the waters of the world cascade down from Mount Olympus, breaking into nine rivers through the earth, the last of which is the River Styx which plummets into Tartarus, perhaps to start the whole cycle afresh. Nine is also the number of years of ostracism meted out as penalty for gods who forswore themselves. When Hesiod uses a number, it's usually nine.

What is nine's significance? One explanation is that ten, the number of fingers on two hands, represents completeness and perfection. Nothing on earth is perfect, and, what with the whims, jealousies, and flirtations of the gods, the same can be said of heaven. Thus the number nine may signify falling short of perfection. There are other explanations. From 1500 BC to 396 AD, the Greeks annually celebrated a nine-day harvest festival, the Eleusinian Mysteries, that initiated the people into secrets of when to plant and reap, the meaning of life, and so on—so the number nine has long been linked to knowledge and wisdom; even today, nine justices sit on the Supreme Court. The gestation period of man is nine months, actually 266 days, which is almost exactly nine periods of the moon (265.5 days). So nine could signify birth or life. Then there are any number of mystical relations among the numbers, such as 9 being 3×3, which may have meant that nine is the harmony of harmonies, or something else equally grand.

Hesiod may be the first person to give an answer to the two questions, how high is heaven? how deep is the underworld? He uses the number nine to describe these dimensions:

> For a brazen anvil falling down from heaven nine nights and days would reach the earth upon the tenth: and again, a brazen anvil falling from earth nine nights and days would reach Tartarus upon the tenth [**40**, p. 131, lines 722–25].

Does Hesiod's description make sense? Is nine days the best measure for both expanses? Imagine yourself as Hesiod's science advisor, a muse of sorts. What words could you whisper into Hesiod's ear, so that what he wrote would ring true down through time until the present?

Here are some developments to keep in mind while formulating an answer.

Not long after Hesiod, the Greeks adopted a series of concentric spheres to model the cosmos. In the sixth century BC, Anaximander asserted that the earth developed from a watery sphere surrounded by fire; the fire misted the waters, creating the atmosphere. Philolaos of the fifth century BC proposed a cosmos of ten concentric spheres, including one for the earth and one for the sun, rotating around a fiery center.

As reported by Plato, in 399 BC, Socrates, in his farewell address before drinking a cup of hemlock, spoke extensively about the inner structure of a round earth, how nine rivers flow through chasms back and forth to earth's center so as to account for the tides. His description is more explicit than Hesiod's. Thoughts of the underworld evidently weighed heavily on Socrates' mind. In a few moments hence he would enter the very realm of which he spoke and see for himself whether his words were true.

Socrates describes the upper boundary of earth's sphere:

> If anyone should come to the top of the air, he could lift his head above it and see, as fishes lift their heads out of the water and see the things in our world, so he would see things in that upper world, things even more superior to those in this world of ours [**62**, *Phaedo*, p. 377].

Socrates assumed that our world is a shadow of that better reality. This leads to the idea that the mechanism of motion may be different in the heavens than on the earth.

Aristotle taught that the earth was at the center of these spheres:

[All] heavy bodies move to the centre of the earth because it [the earth] has its centre at the centre of the universe [5, book ii, chapter ii, p. 245].

Around 275 BC, Aristarchus of Samos proposed that the sun should be at the center of these spheres. Three and a half centuries later, in a remarkable Copernicus-sounding passage from an essay on the man in the moon, Plutarch paraphrased Aristarchus:

It has not been proved that the earth is the center of the sum of things, in all probability things [on the moon] converge upon the moon and remain there [63, *Concerning the face which appears in the orb of the moon*, p. 71].

About 150 AD, Ptolemy hypothesized nine levels of rotating, concentric heavens, one for each of the five planets, the moon, and the sun, with a fixed earth at its center. Figure 2 is a sixteenth century rendition of Ptolemy's

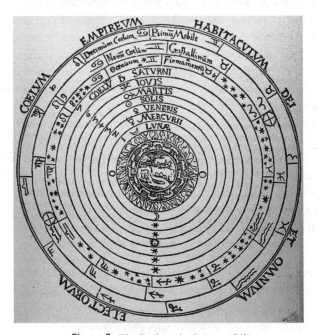

Figure 2. The Ptolemaic Cosmos [4]

cosmos. A tenth and outermost level was reserved for the habitation of God, the unmoveable mover. In accordance with Aristotle, Ptolemy's model assumed that if anything was to fall, it would fall towards earth's center along a straight line. This model, which worked well for astronomical predictions, endured until after the time of Copernicus.

Dante, in the *Divine Comedy*, written in about 1300, describes the greatest of all falls, both in the physical and spiritual sense. Just after Dante and Virgil have climbed down Satan's torso, and while they are climbing up a passage towards Mount Purgatory on earth's surface, Virgil explains,

> And he [Satan] whose pelt our ladder was, stands still
>
> Fixt in the self-same place, and does not stir.
> This side the world from out high Heaven he fell;
> The land which here stood forth fled back dismayed,
> Pulling the sea upon her like a veil,
>
> And sought our hemisphere; with equal dread,
> Belike, that peak of earth which still is found
> This side, rushed up, and so this void was made.
>
> Canto xxxiv, lines 119–24, [**22**, p. 288]

Despite a confusing use of the indeterminate phrase *this side* in the above passage—Virgil could be hand-gesturing as he speaks—we make the following interpretation. When Satan is cast from heaven, he falls head downward towards the earth. Striking the earth in the southern hemisphere, he plows into earth's depths, tunneling out a hole. This induces land to rise above the waters in the northern hemisphere, creating a conical void called Hell in the process, while pushing some of the earth back up into a mountain, Mount Purgatory, in the southern hemisphere, as is depicted in Figure 3. At earth's core, Satan's fall stops. He's forever mired in solid rock and ice, his crotch at the very center of the earth.

Just as the heavens are divided into nine layers, Dante subdivides hell into nine concentric layers of malefaction and adds a tenth, the Vestibule of the Futile, for those with neither faith nor works.

Problems of weight and falling have troubled people through the centuries. In the thirteenth century, Roger Bacon thought about a balance scale. Bacon quotes Aristotle: "The lower a body sinks, the more weight it acquires," [**7**, p. 191]. Call the two arms of the scale A and B. Suppose that the weights on the arms are equal. Press arm A down. Since arm A is nearer to earth's center, it's now heavier than arm B; so why doesn't arm A stay down? And

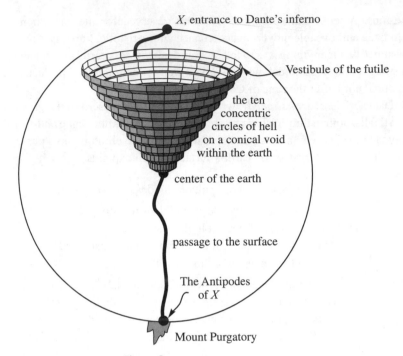

Figure 3. The void of Dante's hell

if arm *A* does rise, why does it proceed upward beyond the equilibrium position? Bacon's argument becomes increasingly convoluted, and he finally concludes that the up and down motion of the scales

> is not due to the nature of the weight itself, but to the impacts of the particles of air [that] retains the motion readily, and therefore for a long time its particles vibrate hither and thither and do not permit the weight to come to rest at once in the position of equality [**7**, p. 194–5].

Although it's unfair to ask this of him, it would be interesting to know how he would have explained why a scale oscillates in a vacuum. His answer might involve the influence of ether, the mysterious fifth building block of the cosmos that supposedly filled otherwise empty places.

In about 1500, Leonardo da Vinci summarized his understanding of why objects fall,

> Every natural action is made in the shortest way: this is why the free descent of the heavy body is made towards the center of the world because it is the shortest space between the movable thing and the lowest depth of the universe [**23**, p. 551].

In the early seventeenth century, Galileo, by experiment, discovered that the distance a body falls is proportional to the square of the time it has been falling. He then extrapolated and used his formula to determine the speed at which a ball falling from the moon would strike the earth. Galileo treated the ball as if it, too, was Lucifer falling from heaven towards the center of the earth.

A contemporary of Galileo, Johannes Kepler, showed that elliptical orbits with the sun at a focus were consistent with the planetary observations accumulated by the Danish astronomer Tycho Brahe. Some kind of phenomenon must keep the planets in their elliptical orbits. But what? One popular theory involved the mechanism of celestial ethereal storms, somewhat like earth's prevailing westerly winds!

In writing the *Principia* of 1687, Newton had no idea why two masses should attract one another. However, given an attractive force between particles of matter proportional to the product of their masses and inversely proportional to the square of their separation, he demonstrated that Kepler's laws of planetary motion follow.

Today everyone knows that objects fall because of gravity and that the stars move in accordance with gravity. Who could doubt it? However—and this is humbling—no one has a clue as to why all things seem to attract one another. They just do! Physicists such as George Gamow have proposed the existence of gravitons, elementary particles that transfer energy in a simple way, and which when taken collectively induce Newton's law of attraction. See Linwood [49] for a readable account of this derivation. Yet, even if gravitons exist, the question remains as to why they do what they do. We may have to conclude that magic is somehow at the root of motion. With such a state of our current understanding, not to mention quantum mechanics and general relativity, the shade of Zeno—the old Greek master who argued that motion is impossible—must be smiling in ironic empathy.

Newton's achievement is an example of the great mystery of the connection between abstraction and reality. Why should mathematical models so faithfully predict the behavior of matter? Eugene Paul Wigner, a Nobel laureate in physics, called this baffling phenomenon the unreasonable effectiveness of mathematics in the natural sciences, [96]. Aristotle made a similar observation:

Mathematical impossibilities will be physical impossibilities too, but this proposition cannot be simply converted, since the method of mathematics is to abstract, but of natural science to add together all determining characteristics [5, book i, chapter i, p. 263].

Abstract mathematics and physical occurrences are radically different. The Pythagoreans bridged this radical gap by believing that nature was constructed of numbers, and Plato by believing that nature was constructed of geometrical forms. But since monads (numbers and points) have no physical properties, Aristotle concluded that "monads in combination cannot either produce bodies or possess weight," [**5**, book i, chapter i, p. 269]. As Aristotle saw it, there is no way for reality to be constructed from abstraction—there is no way for mathematics to "be simply converted" into bodies.

While the connection between the mind's models and matter's phenomena may never be resolved formally, their intuitive connection is compelling.

How so? Figure 4 is a drawing of Winnie the Pooh and his friends, Piglet, Rabbit, and Roo, playing poohsticks. The rules are simple. Each player selects a stick. Players lean over a rail on the upriver side of a bridge. Simultaneously,

Figure 4. Playing poohsticks

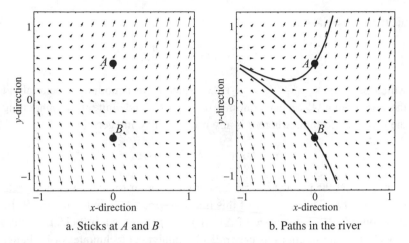

a. Sticks at A and B b. Paths in the river

Figure 5. Playing poohsticks with a differential equation

they drop their sticks into the current. Then they rush to the downriver side
of the bridge. Whoever's stick comes first into view is the winner, [**55**,
p. 237–40].

Pooh and his friends are really playing with differential equations. Consider the differential equation

$$\frac{dy}{dx} = 2x + 3y. \tag{1}$$

Let the region Q be the rectangle $[-1, 1] \times [-1, 1]$ in the xy-plane. Think
of Q as a part of a river's surface. For any point (x, y) in region Q, think
of the current of the river as flowing in the direction given by the slope
$f(x, y) = 2x + 3y$. Figure 5a shows the direction flows at a few points as
little arrows. Now think of dropping in poohsticks at initial points $A = (0, .5)$
and $B = (0, -.5)$. The sticks will flow with the current, tracing out paths
as indicated in Figure 5b. They represent solutions to (1) through points A
and B.

For any sufficiently nice function $f(x, y)$ and any point A, a unique
solution path through A for $y' = f(x, y)$ exists.

Leonhard Euler gave a simple way to approximate these solutions that is
especially useful when analytical techniques fail. Here's the idea. Start with
an initial point $A = (x_0, y_0)$. Choose a small step size Δx in the x-direction,
and let $x_1 = x_0 + \Delta x$. Let y_1 be an approximation to the y-coordinate
corresponding to x_1 through which the poohstick dropped at A will pass.
By the differential equation, we know the direction the poohstick takes

at A, and so write

$$f(x_0, y_0) \approx \frac{y_1 - y_0}{\Delta x},$$

which gives $y_1 \approx y_0 + f(x_0, y_0) \Delta x$. Since there is nothing special about the initial point A, we can continue this process indefinitely, so generating the recursion formula where n is any non-negative integer:

$$y_{n+1} \approx y_n + f(x_n, y_n)\Delta x. \tag{2}$$

Now connect the points (x_n, y_n) in sequential order to give an approximate solution curve. The accuracy of this method depends upon the nature of the function $f(x, y)$ and the size of Δx. If $y' = f(x, y)$ is a sufficiently robust model, then the solutions as generated by analytical techniques or by better variations of (2) are accurate predictors of the phenomenon. We just float on down the dual rivers of abstraction and reality in pleasant accord.

Euler's method works just as well for higher order differential equations. For the second-order differential equation

$$y'' = \frac{d^2 y}{dx^2} = f(x, y, y'), \tag{3}$$

if we let $z = y'$, Euler's method becomes,

$$\begin{cases} y_{n+1} \approx y_n + z_n \, \Delta x, \\ z_{n+1} \approx z_n + f(x_n, y_n, z_n) \, \Delta x. \end{cases} \tag{4}$$

If $y = y_0$ and $y' = z_0$ when $x = x_0$, we have initial conditions to get the method started. This time, the differential equation is not about motion on the surface of a river. The arrows analogous to the two-dimensional ones of Figure 5 are now three-dimensional and the solution to (4) is a path through three space.

Solving second-order differential equations is like playing three-dimensional poohsticks, as exemplified in a story by Chuang Tzu, a Taoist philosopher of the third century BC.

Confucius was seeing the sights at Lü-liang, where the water falls from a height of thirty fathoms and races and boils along for forty li, so swift that no fish or other water creature can swim in it. He saw a man dive into the water. After the man had gone a couple of hundred paces, he came out of the water and began strolling along the base of the embankment.

Confucius ran after him and said, "At first I thought you were a ghost. May I ask if you have some special way of staying afloat?"

[The man replied,] "I go under with the swirls and come out with the eddies, following along the way the water goes." [**19**, p. 126]

The solution curves of second-order differential equations dive and flow, just like the man or poohstick in the story, following the way of the water. Those unfamiliar with calculus might swear that the abstract prediction of reality made by differential equations is magic. Actually, mathematical modeling and reality flow along as one. To achieve this oneness, of drifting along with the current of reality without losing a sense of understanding, we need to find suitable variables and a flow function f to make a model that is simple enough to manipulate, yet complex enough to simulate behavior of the physical action being observed.

As optimists in our ability to understand the universe, we imagine that a contrapositive version of Aristotle's statement about mathematical impossibilities being physical impossibilities is true: that any physical motion can be explained mathematically. For most of this book, (2) and (4) serve as a partial answer with respect to modeling motion. Listening for the muses—whether that be the mathematical masters of the past or simple intuition—we will fill in the details, constructing flow functions, so as to understand falling and spinning. In the last chapter, we give an answer as to advising Hesiod about anvil falling times.

preamble ii

> Towers fall with heavier crash.
>
> Horace (65–8, BC), *Odes*

One morning in calculus class, while discussing the standard problem of a falling body near the surface of the earth, I asked how gravity changes with distance above the surface of the earth. The students replied with understanding, quoting the inverse square law, so I asked a natural reciprocal question. How does gravity change with distance below the surface of the earth?

One student gave the idea that if you go down a hole, the mass of the ground overhead exerts an upward force, so you weigh less. Therefore, gravity decreases with descent.

Such reasoning sounded good. However, I had been certain that gravity increased with descent, thinking as Aristotle had that as an object falls it gets heavier. Now I was confused. I couldn't see any flaw in the argument nor could I see why my original thinking was incorrect. Confessing my ignorance, I promised the class I'd get back to them with a definitive answer.

In a geophysics text, I found a graph similar to Figure 5e of the next chapter, a plot of gravity's strength with respect to distance from earth's center, demonstrating that the relation depends upon the model of the earth being used. Ultimately, the answer for the class grew into an article for the *Mathematics Magazine*.

One of the figures for the first draft had been a map of the earth as drawn by a hollow earth society. I wanted the map to add pizzazz. But how could I ask for permission from them, as my article was nothing but ridicule of them? I was stuck. However, the editor had a solution. He commissioned an artist to design the magazine cover, who sketched the map along with an outline of an Edmond Halley portrait and a rendition of an Edgar Rice Burroughs paperback cover. The final product, Figure 7 on page 28, is almost good enough for a cover of the pulp science fiction magazines of a generation or two ago.

CHAPTER II

The Gravity of Hades

Imagine being inside a chamber deep within the earth, such as depicted in the Jules Verne cavern of Figure 1. Is the acceleration due to gravity stronger or weaker in the chamber than at the surface of the earth? Here's how Leonhard Euler answered the question in 1760:

> We are certain that gravity acts with the greatest force at the surface of the earth, and is diminished in proportion as it removes from thence, whether by penetrating towards the center or rising above the surface of the globe [**28**, vol. i, letter L, p. 182].

Figure 1. Frontispiece for Verne's 1871 publication of *Journey to the Center of the Earth*, 100 miles below Iceland, [**91**]

To answer this question, we consider various models of earth's structure that have been proposed over the years and show

- for homogeneously dense planets—Euler's implicit model—gravity weakens with descent from the surface
- for a planet with a homogeneous mantle that is less than 2/3 as dense as its core, a local minimum for gravity exists if the mantle is sufficiently wide
- for our earth—having a solid inner core, a less dense liquid outer core, and an extensive even less dense mantle—gravity intensifies with descent from -9.8 m/sec^2 at the surface to an extreme of -10.8 m/sec^2 where the mantle meets the core.

We also give a condition involving only surface density and mean density that determines whether gravity increases or decreases with depth from the surface of a planet and we conclude with a body falling down a hole through the earth.

Some preliminary classical mechanics

To find the gravitational acceleration induced by the earth on a particular point-mass P, we follow Newton and first find the gravitational acceleration induced by each point-mass of the earth on the point-mass P, and then sum. Let the mass m of a body be condensed to the single point $Q = (x, y, z)$. We wish to determine the gravitational acceleration $a(s)$ induced by Q in the direction $\mathbf{k} = (0, 0, 1)$ on $P = (0, 0, s)$, as indicated in Figure 2.

The reader unfamiliar with vector calculus may wish to look over the first seven items of the appendix before reading these next two pages.

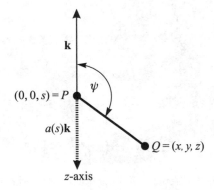

Figure 2. The pull $a(s)$ by Q on P

We start with Newton's law of gravitation, the force of gravity between two point masses a and b separated by r units is Gab/r^2, where $G \approx 6.67 \times 10^{-11}$ Nm²/kg², the universal gravitation constant. Since force on a mass is equal to mass times acceleration, the gravitational acceleration on P in the direction $\overrightarrow{PQ}$ is $Gm/(x^2 + y^2 + (z - s)^2)$. But we want only the portion in the **k** direction. If ψ is the angle between $\overrightarrow{PQ}$ and **k**, then

$$\cos \psi = \frac{(x, y, z - s) \cdot (0, 0, 1)}{\sqrt{x^2 + y^2 + (z - s)^2}} = \frac{z - s}{\sqrt{x^2 + y^2 + (z - s)^2}}.$$

Therefore $a(s) = \cos(\psi)Gm/\left(x^2 + y^2 + (z - s)^2\right)$, which we write as

$$a(s) = \frac{Gm(z - s)}{(x^2 + y^2 + (z - s)^2)^{3/2}}. \qquad (1)$$

Acceleration on a mass at $(0, 0, s)$ by a point mass at (x, y, z)

Ideal planet models

To apply (1) to calculate the gravitational acceleration $a(s)$ for a heavenly body at distance s from its center of mass, we focus on ideal planets, whose concentric spherical shells have constant density. Though earth is best described as an oblate spheroid flattened at the poles by about 14 km and bulging at the equator by about 7 km, and though its shells are not quite of constant density, we view earth as an ideal planet unless stated otherwise. For each ideal planet, we show that $a(s) \leq 0$ for all $s \geq 0$. Therefore, the greatest magnitude for acceleration occurs when $a(s)$ is a minimum.

The simplest ideal planet is a single nonzero dense onion layer, or a soap bubble.

A soap bubble model. Let S be a spherical shell of mass M with constant density, center $O = (0, 0, 0)$ and radius r. Then the gravitational acceleration on P due to the shell S is

$$a(s) = \begin{cases} 0, & \text{if } 0 \leq s < r, \\ -\dfrac{GM}{2s^2}, & \text{if } s = r, \\ -\dfrac{GM}{s^2}, & \text{if } s > r. \end{cases} \qquad (2)$$

Acceleration by a spherical shell

The derivation of (2) can be found in classical mechanics texts, e.g. Fowles and Cassiday [33, pp. 207–9]. We will derive (2) by integrating (1) over the shell S. Let

$$\mathbf{r} = (r \cos \theta \sin \phi, r \sin \theta \sin \phi, r \cos \phi)$$

be a parameterization of S where $0 \leq \theta \leq 2\pi$ and $0 \leq \phi \leq \pi$. Then the area element for S is

$$\left\| \frac{\partial \mathbf{r}}{\partial \theta} \times \frac{\partial \mathbf{r}}{\partial \phi} \right\| d\theta d\phi,$$

or $r^2 \sin \phi d\theta d\phi$. Since the surface area of a sphere of radius r is $4\pi r^2$, the point-mass m of (1) can be taken as

$$M/(4\pi r^2)r^2 \sin \phi d\theta d\phi = M/(4\pi) \sin \phi d\theta d\phi.$$

Therefore from (1), the acceleration $a(s)$ by S on the point-mass at P is

$$a(s) = \int_0^\pi \int_0^{2\pi} \frac{G\frac{M}{4\pi}(r \cos \phi - s) \sin \phi}{(r^2 + s^2 - 2rs \cos \phi)^{\frac{3}{2}}} d\theta d\phi \qquad (3)$$

$$= \int_0^\pi \frac{G\frac{M}{2}(r \cos \phi - s) \sin \phi}{(r^2 + s^2 - 2rs \cos \phi)^{\frac{3}{2}}} d\phi.$$

The change of variable $u = \cos \phi$ gives

$$a(s) = \frac{GM}{2} \int_{-1}^1 \frac{ru - s}{(r^2 + s^2 - 2rsu)^{\frac{3}{2}}} du, \qquad (4)$$

$$a(s) = \frac{GM}{2s^2} \left(\frac{r - s}{\sqrt{(r - s)^2}} - \frac{r + s}{\sqrt{(r + s)^2}} \right). \qquad (5)$$

If $0 < s < r$ then (5) gives 0; if $s > r$ then (5) gives $-GM/s^2$. To obtain the result for $s = r$, simplify (4) before integrating. Since a general ideal planet is a series of concentric soap bubbles, when s exceeds R, (2) gives the familiar principle that when dealing with forces exerted by heavenly bodies their masses can be treated as point masses at their centers. When $0 \leq s < R$, (2) is a well-known result of electrostatics, which has features in common with gravitation. That is, for a hollow metal sphere carrying a surface charge, the electrical field within the sphere is 0. However, for any void within a charged metal object, the field is also 0 [31, chapter 5], which is not the case for the gravitational field within an arbitrary void of a planet.

As an example showing that neither of these two properties necessarily hold if a heavenly body is not an ideal planet, let S be a hollow right circular cylinder with mass M, height 2, radius 1, center at the origin, and central axis parallel to the z-axis. Since the surface area of S is 6π, assume that S has uniform density of $M/(6\pi)$. To calculate the gravitational acceleration induced by S on the point mass at $(0, 0, s)$ where $s \geq 0$, use (1) and integrate over the top, the bottom, and the sides of S, giving $a(s)$ as

$$\frac{GM}{6\pi} \left(\int_0^1 \int_0^{2\pi} \frac{(1-s)r}{(r^2 + (1-s)^2)^{\frac{3}{2}}} d\theta\,dr - \int_0^1 \int_0^{2\pi} \frac{(1+s)r}{(r^2 + (1+s)^2)^{\frac{3}{2}}} d\theta\,dr \right.$$

$$\left. + \int_{-1}^1 \int_0^{2\pi} \frac{(z-s)}{(1 + (z-s)^2)^{\frac{3}{2}}} d\theta\,dz \right).$$

This becomes

$$a(s) = \begin{cases} \dfrac{GM}{3} \left(\dfrac{s+2}{\sqrt{1 + (1+s)^2}} + \dfrac{s-2}{\sqrt{1 + (1-s)^2}} \right), & \text{if } 0 \leq s < 1, \\[2em] \dfrac{GM(3 - 2\sqrt{5})}{15}, & \text{if } s = 1, \\[2em] \dfrac{GM}{3} \left(\dfrac{s+2}{\sqrt{1 + (1+s)^2}} + \dfrac{s-2}{\sqrt{1 + (1-s)^2}} - 2 \right), & \text{if } s > 1. \end{cases} \qquad (6)$$

For $0 < s < 1$, $a(s) > 0$, which means that if a particle is on the z-axis between 0 and 1, then it is attracted towards the top of the cylinder. As $s \to \infty$, $s^2 a(s)$ is a constant, so $a(s)$ is asymptotically an inverse square law.

In general, when calculating the gravitational acceleration induced by a heavenly body on a certain point-mass, any nonspherical symmetry of the body produces nightmarish integrals. But fortunately for astrophysics, planets and stars are more or less ideal planets. Also, many of the central problems of celestial mechanics involve forces of gravitation at many radii from the heavenly bodies in question, so that the inverse square law can be used with confidence and relief.

To find the cumulative gravitational acceleration on a point-mass as induced by the point-masses of a nontrivial ideal planet, it is convenient to sum them layer by layer. With this idea in mind, we integrate (2) appropriately to find $a(s)$.

The general density model.   Let S be a ball of radius R, mass M, and center O whose density at s units from O is given by $\delta(s)$. That is, S is radially symmetric with respect to its density. Then

$$
a(s) = \begin{cases} -\dfrac{4\pi G}{s^2} \displaystyle\int_0^s \rho^2 \delta(\rho) d\rho, & \text{if } 0 \leq s \leq R, \\[2em] -\dfrac{GM}{s^2}, & \text{if } s \geq R. \end{cases} \tag{7}
$$

Acceleration by an ideal planet with general density

To derive (7) from (2) we note that the mass of the shell at radius ρ is $4\pi\rho^2\delta(\rho)$. Therefore,

$$
a(s) = \begin{cases} -\displaystyle\int_0^s \dfrac{4G\pi\rho^2\delta(\rho)}{s^2} d\rho, & \text{if } 0 \leq s \leq R, \\[2em] -\displaystyle\int_0^R \dfrac{4G\pi\rho^2\delta(\rho)}{s^2} d\rho & \text{if } s \geq R. \end{cases}
$$

Since $\int_0^R 4\pi\rho^2\delta(\rho)d\rho = M$, this gives (7).

Constant acceleration: an old implicit model

When reading of mythical heroes venturing off into the underworld, such as Orpheus seeking to reclaim Eurydice or Hercules throttling Cerebus, it seems that no matter how deep the heroes descend into the earth, gravity remains constant. Here is a snippet from Ovid's *Metamorphoses*, describing Orpheus and Eurydice's ascent from Hades:

> Now thro' the noiseless throng their way they bend,
> And both with pain the rugged road ascend;
> Dark was the path, and difficult, and steep,
> And thick with vapours from the smoky deep [**61**, book x].

The difficulty seems to arise from smoke and steepness, not increased gravity. The sketches in Figure 3, based on two old Grecian urns, show the two heroes in Hades. Orpheus is serenading the lords of the underworld and Hercules is about to chain Cerberus. Turner [**88**] enumerates other literary characters who frequented the underworld.

Figure 3. Orpheus and Hercules in Hades

In *The Divine Comedy*, written about 1300 AD, Dante journeys with his ghost guide Virgil from the earth's surface, down through ten concentric levels of Hell, to the very center of the earth. Figure 4 loosely labels the categories in Dante's model. When Dante scrambles over a broken bridge deep within Hell, it is with the same effort he would have used along a road to Rome. Arrows and other projectiles in Hell appear to follow the same trajectories as at the surface of the earth. When Dante clambers about the hip of the gargantuan and hair-covered Satan at the literal center of the earth on his way to the antipodes, he changes directions, maneuvering with a degree of exertion like that of reversing his position while clinging to a root-covered, vertical rockface in the Apennines.

> And when we had come to where the huge thigh-bone
>> Rides in its socket at the haunch's swell,
>> My guide, with labour and great exertion,
>
> Turned head to where his feet had been, and fell
>> To hoisting himself up upon the hair,
>> So that I thought us mounting back to Hell.
>>> Canto xxxiv, lines 76–81 [**22**, p. 287]

Up and down reverse dramatically. Dante's Satan (more precisely, Satan's crotch) is a veritable singularity. After climbing on a bit more, Dante takes a rest, looks back, and sees Lucifer's legs sticking up.

> And if I stood dumbfounded and aghast,
>> Let those thick-witted gentry judge and say,
>> Who do not see what point it was I'd passed.
>>> Canto xxxiv, lines 91–93 [**22**, p. 287]

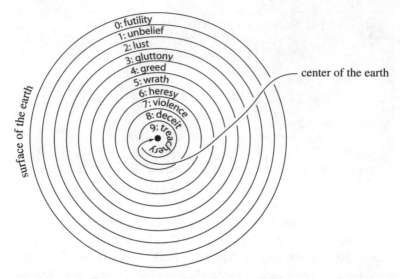

Figure 4. Concentric circles in Dante's inferno

Dante may have borrowed this image from the first century writer, Plutarch, who mentions this same singularity, albeit less graphically so:

> If a man should so coalesce with the earth that its center is at his navel, the same person at the same time has his head up and his feet up too [**63**, p. 67].

This literary evidence suggests that the implicit density model of the earth used in these imaginings is that the density function $\delta(s)$ generates a constant gravitational acceleration. For $0 < s < R$, such a density function satisfies

$$-\frac{4\pi G}{s^2} \int_0^s \rho^2 \delta(\rho)d\rho = C,$$

where C is some constant. Multiplying by s^2 and differentiating yields $\delta(s) = k/s$, where k is a constant. That is, the ancient model of the earth appears to be one wherein density and distance from the earth's center vary inversely! Even in such a model, where the density near the center is unbounded, the mass of the earth is finite.

A common-notion constant density model

Insisting that gravity remain constant on descent into the earth requires a singularity in its density. Much simpler and more natural is to take the density as constant. In an informal survey of several hundred colleagues both in mathematics and physics which I conducted a few years ago, their almost universal and emphatic answer, "Gravity decreases with descent!" subsumes this model. Perhaps one of the reasons for such ardency from physicists and mathematicians is that almost all of us in our undergraduate differential equation days solved an exercise very much like this one, (which we solve in general at the end of this chapter):

> Inside the earth, the force of gravity is proportional to the distance from the center. If a hole is drilled through the earth from pole to pole, and a rock is dropped in the hole, with what velocity will it reach the center? [72, p. 24]

A tempting reason to guess that gravity decreases with depth is the following argument, appealing to intuition, but incorrect: "As we go down, the portion of the earth above will exert an upward force on us, hence lessening the downward force exerted by the remainder of the earth." But by (2), at s units from the center, only the mass within s units of the center determines $a(s)$.

Taking a ball S of uniform density δ, mass M, and radius R, gives, using (6),

$$a(s) = \begin{cases} -\dfrac{4\pi G \delta s}{3}, & \text{if } 0 \leq s \leq R, \\[2mm] -\dfrac{GM}{s^2} & \text{if } s \geq R. \end{cases} \tag{8}$$

Acceleration by a planet of constant density

That is, gravity and distance from the center are directly proportional up to the surface of the earth, as shown in Figure 5a.

The reader may enjoy contrasting (8) with the acceleration functions of non-ideal planets of constant density. For example, let S be a right circular cylinder of radius 1, height 2, mass M, center at the origin, and central axis parallel to the z-axis. Integrating (1) over S gives the gravitational acceleration $a(s)$ on a mass at $(0, 0, s)$ induced by S for $s > 0$ as

$$a(s) = MG\left(\sqrt{1 + (1 + s)^2} - \sqrt{1 + (1 - s)^2} + |1 - s| - |1 + s|\right),$$

whose graph looks much like Figure 5a. This shows that the seemingly artificial limit $\lim_{x \to \infty} \sqrt{1 + (1 + x)^2} - \sqrt{1 + (1 - x)^2} = 2$ that appears in many texts actually arises in a physical situation.

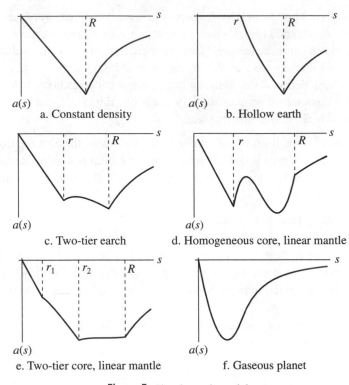

Figure 5. Simple earth models

Halley's hollow earth model

Edmond Halley, the "Father of Geophysics," viewed Newton's (erroneous) calculation in 1687 of the relative densities of the earth and the moon being 5 to 9 as one of the more significant of the discoveries presented in the *Principia*. Believing Newton to be correct and yet wanting the material of the earth and moon to be equally dense, Halley proposed that the earth is 4/9 hollow. Since his study of magnetic compass variations suggested that the earth had four separate north poles, he went on to claim that the earth's outer shell was 500 miles thick and that inside were three more concentric shells, being the radii of Venus, Mars, and Mercury. Figure 6 shows Halley, in a 1736 portrait that hangs in the halls of the Royal Society, holding his hollow earth sketch. To explain the aurora borealis phenomenon, he postulated that the atmospheres over each of the shells were alive with magnetic lightning-like flashes and that life probably existed on their inner surfaces [45], [46].

Figure 6. Halley at 80 with sketch of hollow earth model, portrait by Thomas Murray, courtesy of the Royal Society

One legacy of Newton's error is hollow earth societies that apparently love to speak of UFOs and superior creatures living inside our planet. A web page [**83**] contains a map of this inner world, shown in an illustration by Jason Challas on the cover of the *Mathematics Magazine*, Figure 7.

A well-known science fiction tale inspired by Halley's model is by Tarzan's creator, Edgar Rice Burroughs. In *At the Earth's Core*, written in 1908, two heroes in an out-of-control tunneling machine burrow from the surface of the earth and emerge three days later and 500 miles down in the inner world of Pellucidar. Figure 8 is a sketch of a book cover depicting this earth ship, looking not too different from the earth-eating machines that dug the tunnel under the English Channel. At the half-way point, 250 miles deep, up and down switch for Burroughs' characters in the same way as when Dante climbed down Satan's torso at the center of the earth. Is it possible to walk upright on the inside of the earth's outer layer?

Let S be a shell of inner radius r and outer radius R, uniform density δ, and mass M. Then (7) gives,

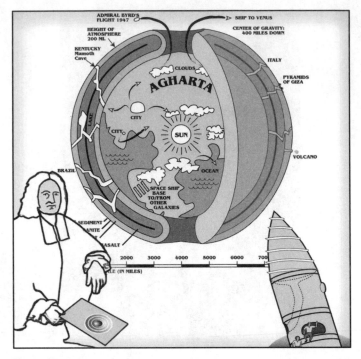

Figure 7. Map of a hollow earth, [**75**, cover], courtesy of Jason Challas

$$
a(s) = \begin{cases}
0, & \text{if } 0 \leq s \leq r, \\[2mm]
-\dfrac{4}{3}\pi G\delta\left(s - \dfrac{r^3}{s^2}\right), & \text{if } r \leq s \leq R, \\[2mm]
-\dfrac{GM}{s^2} & \text{if } s \geq R.
\end{cases}
\tag{9}
$$

Acceleration by a planet with a hollow core

Figure 5b illustrates this acceleration. The entire empty core is a zero-gravity haven.

Burroughs does have his character sense "a certain airy lightness of step" on Pellucidar, and explains, "The force of gravity is less upon the inner world due to the counter-attraction of that portion of the earth directly opposite the spot at which one's calculations are made" [**16**, *Freedom* chapter], which is probably how the typical man-in-the-street might reason.

Figure 8. On Pellucidar

Rudy Rucker, a mathematician and novelist, in a historical fantasy based upon an incident in 1842 when the U.S. Congress actually funded a scientific expedition to explore the hollow earth, describes what life might be like living in a Pellucidar-type world in which the only gravity is that induced by the spin of the earth. Rucker's narrator describes a hike through a jungle on this world:

> Our progress was easy. In the all but weightless surroundings, we could hop from branch to branch like squirrels—though it took some bone-shaking crashes till we learned not to jump too hard. It was so strange to be living with practically no gravity. How odd and how wonderful that the attraction of the great domed planet crust behind us exactly balanced the pull of the land close ahead [**68**, p. 167].

Hooke's onion model

Robert Hooke posited an earth with multiple layers, one reason for which was to explain why the magnetic north pole appears to wander. His idea was that the magnetic source might be embedded in a layer that rotated at a rate slightly different than the surface layer. As an example of this model, suppose that the earth consists of two homogeneous layers, the inner one called the core and the outer one called the mantle. Let S be a ball consisting of an inner core of density δ_1, mass M_1, radius r, and of an outer mantle, density δ_2, mass M_2, inner radius r and outer radius R. Two applications of (7) yield

$$a(s) = \begin{cases} -\dfrac{4\pi G\delta_1 s}{3}, & \text{if } 0 \le s \le r, \\[2mm] -\dfrac{GM_1}{s^2} - \dfrac{4}{3}\pi G\delta_2 \left(s - \dfrac{r^3}{s^2} \right), & \text{if } r \le s \le R, \\[2mm] -\dfrac{G(M_1 + M_2)}{s^2}, & \text{if } s \ge R. \end{cases} \qquad (10)$$

Acceleration by an onion model with two layers

The mantle contains a local extreme for gravitational acceleration at $s_c = r\sqrt[3]{2(\delta_1 - \delta_2)/\delta_2}$ provided $r < s_c < R$, so a necessary condition for the existence of a local extreme is $\delta_2/\delta_1 < 2/3$. That is, a planet's mantle must be no more than $2/3$ as dense as its core in order for an extreme to exist. Figure 5c illustrates the acceleration function when $\delta_1 = 2\delta_2$ and $R = 2r$.

A molten core, variable density mantle model

To account for volcanic activity, geophysicists of the nineteenth century postulated that earth's central region was molten, which suggested that the core might be homogeneous. The solid mantle was presumed to have a density that increased with depth. As an example of this, let S be an ideal planet with a homogeneous core of density δ, mass M_1, radius r, and a variable mantle of density $\delta(s) = \mu + \lambda s$ (where $r < s < R$), and mass M_2. Then

$$a(s) = \begin{cases} -\frac{4}{3}\pi G\delta s, & \text{if } 0 \le s < r, \\[2mm] -\frac{GM_1}{s^2} - 4\pi G\left(\frac{\mu s}{3} + \frac{\lambda s^2}{4} - \mu\frac{r^3}{3s^2} - \lambda\frac{r^4}{4s^2} \right), & \text{if } r \le s \le R, \\[2mm] -\frac{G(M_1+M_2)}{s^2}, & \text{if } s \ge R. \end{cases} \qquad (11)$$

Acceleration by a homogeneous core with mantle of variable density

Figure 5d (not drawn to scale) shows the case when $\delta = 1$, $\mu = 2$, $\lambda = -1$, $r = .82$, and $R = 1.2$. It is surprising that $a(s)$ can display such exotic behavior when $\delta(s)$ is so tamely monotonic.

A seismic model

From earthquake analysis and from the understanding that seismic waves are transmitted differently in liquids and solids, today we think that the earth consists of three layers. With the density unit as 1000 kg/m^3, we'll assume that the earth consists of a solid central core of uniform density $\delta_1 = 13$ and radius 1275 km, a liquid outer core of uniform density $\delta_2 = 10$ and thickness 2225 km, and an affinely dense mantle of thickness 2900 km, with density varying from 6 to 3.3 at the surface. We ignore the thin crust and tenuous atmosphere. Let $r_1 = 1.275$ units (where a unit is 1000 km), $r_2 = 3.5$, and $R = 6.4$, and let M_1, M_2, and M_3 be the masses of the inner core, the outer core, and the mantle. Then within the mantle, $\delta(s) = \mu + \lambda s \approx 9.26 - 0.931s$, where s is in thousands of kilometers, and

$$
a(s) = \begin{cases} -\dfrac{4}{3}\pi G\delta_1 s, & \text{if } 0 \le s < r_1, \\[2ex] -\dfrac{GM_1}{s^2} - \dfrac{4}{3}G\delta_2 \left(s - \dfrac{r_1^3}{s^2} \right), & \text{if } r_1 \le s \le r_2, \\[2ex] -\dfrac{G(M_1 + M_2)}{s^2} \\ \quad -\dfrac{4}{3}\pi G\left(\mu s + 0.75\lambda s^2 - \mu\dfrac{r_2^3}{s^2} - 0.75\lambda\dfrac{r_2^4}{s^2} \right), & \text{if } r_2 \le s \le R, \\[2ex] -\dfrac{G(M_1 + M_2 + M_3)}{s^2}, & \text{if } s \ge R. \end{cases}
\tag{12}
$$

Acceleration by a solid center, molten outer core, variable mantle

Figure 5e shows a graph of $a(s)$. This acceleration function has no local extreme within the outer core because $\delta_2/\delta_1 = 10/13 > 2/3$, a violation of the necessary condition in Hooke's model. The overall extreme value of gravitational acceleration for this model of the earth is approximately -10.2 m/sec^2 at the boundary of the outer core and the mantle. The graph of Figure 5e supports the intuition of those who spun the Greek legends, for Hades couldn't have been much more than 1000 km below Greece, and the acceleration of gravity within this level of the earth is approximately constant!

Studies from 1989 on the interplay between data and models give a density model as in Figure 9, which in turn gives an acceleration curve qualitatively much like Figure 5e; the current best guess of gravity's extreme within the earth is -10.8 m/sec^2 at the core-mantle boundary [51, p. 155].

A gaseous model

To complete a list of planetary models, let us imagine a gas ball of uniform temperature and uniform composition. The density of such a ball is

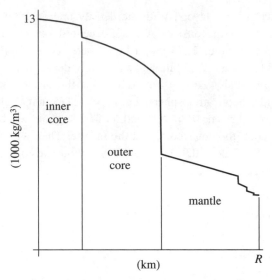

Figure 9. Earth's density, km vs 1000 kg/m^3

exponential [**99**, p. 68]; so $\delta(s) = \delta_o e^{-ks}$ where δ_o is the density at the center, k is a positive constant and $0 \leq s < R = \infty$. Then by (7) for positive s,

$$a(s) = -4\pi G\delta_o \frac{e^{-ks}(k^2 s^2 + 2ks + 2) - 2}{k^3 s^2}. \tag{13}$$

Acceleration by a gaseous planet

Figure 5f shows a typical acceleration graph for this model. The extreme for gravity occurs at $s_c \approx 1.45123/k$, at which point the mass of the ball of radius s_c is $4.50\delta_o/k^3$. (Over 98% of the mass is contained within a ball of radius $3s_c$.) More realistic density models for gas balls are complex and lead to elaborate models for the stars.

A critical condition

As we have seen, given the global density function for an ideal planet, integration tells us how gravity changes with descent. However, even without knowing the global density function, we can determine whether gravity increases or decreases with descent from a planet's surface. All that is needed is the mean density of the planet and its surface density. The surface density of the earth is easy enough to compute—just dig up enough surface rock. But determining the mean density of the earth is more difficult. In Euler's day, the product GM was known, but not its component parts. It was not until 1798, fifteen years after Euler's death, that Henry Cavendish experimentally determined G, which then gave M and hence the mean density of the earth. The following theorem, foreshadowed by the 2/3 ratio for extrema in Hooke's model, shows how local density data can answer the question of how gravity changes with descent. Let P be a planet having radius R and density $\delta(s)$, $0 \leq s \leq R$. Let $m(s)$ be the mass of P up to radial distance s; that is, $m(s) = 4\pi \int_0^s \rho^2 \delta(\rho) d\rho$. Define the mean density $\hat{\delta}(s)$ of P at s as $m(s)$ divided by the volume of a sphere of radius s; that is, $\hat{\delta}(s) = (3/s^3) \int_0^s \rho^2 \delta(\rho) d\rho$. Define the normalized density $\Delta(s)$ at s as $\Delta(s) = \delta(s)/\hat{\delta}(s)$. For example, if $\delta(s)$ is constant, then P's normalized density is $\Delta(s) = 1$. As a second example, if $\delta(s) = k/s$, where k is some constant, as in the constant acceleration model, then P's normalized density is $\Delta(s) = 2/3$.

Theorem 1 (the critical condition) *If $a(s)$ is differentiable, then gravity intensifies with descent whenever $\Delta(s) < 2/3$ and wanes with descent whenever $\Delta(s) > 2/3$. If $\Delta(s) = 2/3$, gravity remains constant with descent.*

To verify this, write (6) for $0 \leq s \leq R$ as

$$-4\pi G \int_0^s \rho^2 \delta(\rho) d\rho = s^2 a(s).$$

By the fundamental theorem of calculus,

$$-4\pi G s^2 \delta(s) = 2sa(s) + s^2 a'(s),$$

so

$$a'(s) = -4\pi G \delta(s) - \frac{2a(s)}{s}. \tag{14}$$

Gravity intensifies with descent precisely when $a'(s) > 0$.

This is equivalent, using (6) and (13), to

$$\frac{-4\pi G}{s^2} \int_0^s \rho^2 \delta(\rho) d\rho = a(s) < -2\pi G s \delta(s),$$

or $(3/s^3) \int_0^s \rho^2 \delta(\rho) d\rho > (3/2)\delta(s)$. That is, $\hat{\delta}(s) > (3/2)\delta(s)$, which says $\Delta(s) < 2/3$. Similarly, if $a(s)$ wanes with descent then $\Delta(s) > 2/3$.

Since the mean density of earth is about $\hat{\delta}(R) = 5515$ kg/m^3 and since crustal rock density $\delta(R)$ is only about half that value, then $\Delta(R) \approx \frac{1}{2}$. This

implies that gravity intensifies with descent from the earth's surface. If the
several thousand moon rock specimens brought back to earth are accurate
indicators, the surface density of the moon is comparable to that of earth.
Since the moon's mean density is about 3330 kg/m^3, $\Delta(R) \approx 0.75$, where
R is the moon's radius, so gravity weakens with descent from the moon's
surface.

Falling through the earth

We now resolve the problem of falling through a hole in the earth:

> If a hole is drilled through the earth from pole to pole, and a rock is
> dropped in it, with what velocity will the rock reach the center, and how
> long will it take?

Plutarch posed a version of this problem when he has a debater list various
reasons as to why some people chose not to believe in a spherical earth:

> Not that masses falling through the earth stop when they arrive at the
> center, though nothing encounter or support them; and, if in their down-
> ward motion the impetus should carry them past the center, they swing
> back again and return of themselves? [**63**, p. 65]

Euler considered this problem as well, explaining in his popular letters to a
German princess that

> You will remember how Voltaire used to laugh at the idea of a hole
> reaching to the centre of the earth, but there is no harm in supposing it,
> in order to discover what would be the end result [**28**, p. 178].

To solve this problem, we neglect air resistance and any relativistic concerns. Let s, v, a, and t be the distance of the rock from the earth's center,
the velocity of the rock, the acceleration of the rock, and time, respectively,
where at $t = 0$, $s = R$, the radius of the earth, $v = 0$, and $a = -g$, where
$g = 9.8$ m/sec^2. Acceleration times velocity can be written as

$$a \frac{ds}{dt} = v \frac{dv}{dt}. \tag{15}$$

Integrating (15) as time t goes from 0 to τ gives

$$\int_0^\tau a \frac{ds}{dt} dt = \int_0^\tau v \frac{dv}{dt} dt. \tag{16}$$

Since both s and v are monotonic functions of t as the rock falls from the earth's surface to its center and since the initial velocity of the rock is 0, (16) becomes

$$\frac{v^2}{2} = \int_R^s adr,$$

where r is a dummy variable representing distance from the earth's center. Since the rock's velocity during fall is negative,

$$v(s) = -\sqrt{2\int_R^s adr}. \tag{17}$$

Thus the speed of the rock as it passes through the earth's center is $v(0)$. To find a formula for the time T it takes for the rock to reach the center, partition the interval $[0, R]$ with $s_0, s_1, \ldots, s_n$ so that $0 = s_0 < s_1 < \cdots < s_n = R$, and $\Delta s = R/n = s_i - s_{i-1}$ for each i, $1 \le i \le n$. Approximate the time for the rock to fall from s_i to s_{i-1} by $\Delta s/(-v_i)$ where $v_i = v(s_{i-1})$ for each i, $1 \le i \le n$. The approximate time T is therefore $\sum_{i=1}^n \Delta s/(-v_i)$. That is,

$$T = \int_0^R \frac{1}{-v} ds. \tag{18}$$

Case 1. The homogeneous earth model. By (8), $a(s) = ks$, for $0 \le s \le R$, where k is a constant. Since $a(R) = -g$, $k = -g/R$. Therefore by (17),

$$v(s) = -\sqrt{2\int_R^s \frac{-g}{R}rdr} = -\sqrt{\frac{g}{R}(R^2 - s^2)}.$$

Thus $v(0) = -\sqrt{gR} \approx 7.9$ km/sec. By (18),

$$T = \int_0^R \frac{1}{-v}dv = \int_0^R \sqrt{\frac{R}{g}}\frac{1}{\sqrt{R^2 - s^2}}ds = \frac{\pi}{2}\sqrt{\frac{R}{g}} \approx 21.2 \text{ minutes.}$$

Case 2. The constant acceleration model. If $a(s) = -g$ for $0 \le s \le R$, then

$$v(s) = -\sqrt{2\int_R^s -gdr} = -\sqrt{2g(R - s)},$$

so that $v(0) = -\sqrt{2gR} \approx 11.2$ km/sec (which also happens to be the escape velocity of the earth), and

$$T = \int_0^R \frac{1}{\sqrt{2g}} \frac{1}{\sqrt{R-s}} ds = \sqrt{\frac{2R}{g}} \approx 19.0 \text{ minutes.}$$

Case 3. An extreme black hole model. We now assume that all the earth's mass is concentrated at its center, so that $a(s) = k/s^2$ by Newton's law of gravitation, for $0 < s < R$. As before, since $a(R) = -g$, $k = -gR^2$. Thus

$$v(s) = -\sqrt{2 \int_R^s \frac{-gR^2}{r^2} dr} = -\sqrt{2gR \left(\frac{R}{s} - 1\right)},$$

so that $v(0) = -\infty$. The time it takes for the rock to reach this singularity, ignoring relativistic considerations, is therefore

$$T = \int_0^R \frac{1}{\sqrt{2gR}} \sqrt{\frac{s}{R-s}} ds = \frac{\pi}{2\sqrt{2}} \sqrt{\frac{R}{g}} \approx 15.0 \text{ minutes.}$$

The greatest falling speed in least time occurs if the acceleration on the rock is always as extreme as is possible, which certainly happens if all the mass of the earth is always nearer to the center of the earth than is the rock. The slowest falling speed and the greatest falling time occur when all the mass is on a shell of radius R, in which case the rock is stationary.

Case 4. The earth. Using (12), (17), and (18) and approximating the integrals gives a speed of $v(0) \approx 8.9$ km/sec and a time $T \approx 19.2$ minutes.

Lucifer cast down. We close with a cosmic case of the previous four examples. In John Milton's *Paradise Lost*, published in 1667, Satan is cast out of Heaven and falls for nine days before landing in Hell.

> Him [Satan] the Almighty Power
> Hurled headlong flaming from th' ethereal sky,
> With hideous ruin and combustion, down
> To bottomless perdition.
>
> Book i, lines 44–47 [56, p. 9]

Although Milton used a Ptolemaic model of the universe in his poem, we will recast the problem as dropping a rock through a galaxy. Let us take the galaxy as the Milky Way, which has a mass M of about 3×10^{41} kg and whose radius R is about 9.3×10^{20} meters. Assume as a first approximation that the Milky Way is a flat, homogeneous pancake, positioned on the xy-plane with its center at the origin. Assume that the rock falls from rest at $(0, 0, R)$. Integrating (1) over the Milky Way gives

$$a(s) = \frac{GM}{\pi R^2} \int_0^R \int_0^{2\pi} \frac{-rs}{(r^2 + s^2)^{\frac{3}{2}}} d\theta dr = \frac{2GM}{R^2} \left(\frac{s}{\sqrt{R^2 + s^2}} - 1 \right).$$

By (17),

$$v(s) = -\frac{2\sqrt{GM}}{R} \sqrt{\sqrt{R^2 + s^2} - s + R(1 - \sqrt{2})}.$$

So $v(0)$ is approximately 225 km/sec. By (18), the rock will reach the center of the milky way in about 340 million years, while light would make this journey in about one hundred thousand years.

Summary and a look ahead

In this chapter we considered various models of the earth, generated their corresponding gravitational acceleration functions, and then determined how rocks fall in them. In Chapter VIII we return to this question and characterize the general properties of gravitational acceleration functions for any ideal planet. We will also take into account the spin of the earth, and see that the paths of rocks falling through a rotating planet are stranger than might be expected.

Exercises

1. Imagine planet U is a homogeneous solid, a right cylinder, with diameter of the base equal to its height, whose mass is that of the earth. Find the gravitational acceleration $a_1(s)$ function along the central axis of U. (Anaximander (611–547 BC) conjectured that the earth was a cylindrical stone pillar whose height was one third the diameter of the base [**39**, p. 10].)

2. For planet U of Exercise 1, determine the gravitational acceleration functions $a_2(s)$ and $a_3(s)$ along a line through U's center which is perpendicular to its central axis and along a line through the rim of U's base and its center.

3. On planet U of Exercise 1, if rocks are dropped down holes simultaneously on the surface using a_1, a_2, and a_3 of the previous two questions, which rock reaches U's center first?

4. Imagine that planet X is a disk whose thickness is the diameter of earth and that it has uniform density equal to the average density of earth. What does the radius of X need to be so that the gravitational acceleration of a person standing on X's surface along the cylinder's axis is -9.8 ft/sec^2?

5. In *The Hollow Earth* [**68**], people living on a Pellucidar-type world experience a mild gravity due to the rotation of the earth. Assume that their atmosphere is a sphere of radius 5600 km, homogeneously dense throughout. Determine the total mass of the atmosphere. What gravitational acceleration will it induce on the inhabitants of Pellucidar?

6. To continue with Exercise 5, suppose that the atmosphere within Pellucidar follows the model of (13). At radial distance 5600 km, assume that the density of the atmosphere is that of sea level. Determine the total mass of the atmosphere and re-calculate the gravitational acceleration for the inhabitants of Pellucidar.

7. Generate a function that approximates the density curve in Figure 9. Use it to generate $a(s)$, the gravitational acceleration s units from the earth's center. With this acceleration function, calculate how long it takes for a rock to fall from the earth's surface to its center. Compare your answer with the results of Case 4, p. 36.

8. Since the episode of Satan's expulsion from heaven described by Milton took place when the galaxy was in chaos, and since Satan was hurled out of heaven, experiment with the pancake model of the universe, the rock's initial velocity and distance from the center, so that Satan's trip can be made in nine days.

9. Imagine that Planet Y is a cube of side length 100 km. Design a density function so that gravity is constant along its surface.

10. Imagine that Planet Z is a point-mass and that a rock is one meter from Z. From rest, allow the rock to fall towards Z. Calculate the mass of Z so that the rock's speed at half of a meter is half that of the speed of light.

preamble iii

Humpty Dumpty sat on a wall.
Humpty Dumpty had a great fall.

anonymous nursery rhyme

The word *ballistics* has dark undertones, evoking shades of violence and war. From tossing stones by hand or sling, to launching arrows with bows, to firing bullets and shells, history drips with blood. The greatest of the ancient mathematicians, Archimedes, counseled his king how to catapult stone weighing a quarter ton into Roman warships [**10**, p. 34]. Some of the best mathematicians and physicists of the last century worked on the atom bomb, which when dropped ended World War II. For better or for worse, mathematicians have advanced the technology of destruction.

The Vietnam War was at its bloodiest when I entered the university. Early in the spring term of my freshmen year all American males born in 1953 had their birth dates assigned a number from 1 to 365 by lottery, to be announced over the radio. Many of us anxiously tuned in. If we were assigned a low number, what would we do? Enlist immediately, wait to be drafted, or do something else? Most of us were undecided. A friend and classmate couldn't bear the suspense of listening, and took a long walk instead, asking me to tell him his number. When my number 322 came up, I felt both elation and guilt, elation because I knew I wouldn't have to shoot at someone shooting at me, and guilt because I knew that some of my friends would soon be asked to do so in my place. I continued listening. My friend's number came up. It was 2. He took the news without flinching. He enlisted—and died in a rice paddy.

Walt Whitman's quote from the first preamble of this book, "I also say it is good to fall," is cryptic out of context. In context, Whitman speaks of the soldier going off to war for civic duty, for honor, for defending freedom—all the noble reasons. To the soldier returning in triumph we give parades and speeches. If it is good not to fall, then for some of the very same reasons it is also good to fall. I like to think that my friend's life was not wasted.

Future years may see more horrific weaponry, all more or less within the realm of ballistics. The Polish science fiction writer, Stanislaw Lem, in stories which slipped by his cold war censors, graphically describes some of these weapons, ones surpassing the smart bomb; read *The Invincible* for example [**47**]. It will be we mathematicians who develop such things. God help us all.

CHAPTER III

Ballistics

Throw a ball into the air. How high and far will it go before falling back to the earth?

One obvious way to answer the question is to load a cannon with an amount of powder, prop the muzzle at an angle, light the fuse, see what happens, and then repeat, and repeat, and repeat with different amounts of powder and angles of elevation. An easier way is to understand the mechanics of the cannonball shot and so determine the relation between time t, distance $s(t)$ of the ball from the cannon, and the speed $v(t)$ of the cannonball.

One of the first to try this easy method was Leonardo da Vinci. A century later the rule was more clearly formalized by Galileo. In another half century, Newton generalized the idea to planetary motion. In this chapter we outline these approaches and then analyze a nineteenth century ballistics shot to the moon.

da Vinci's Solution

Leonardo da Vinci is the prototypical Renaissance man. He was a master of almost everything. He kept notebooks of his various experiments, observations, thoughts, stories, sketches, all totaling thousands of manuscript pages. In one of his notebooks he describes his work:

> This will be a collection without order, made up of many sheets, hoping afterwards to arrange them. I believe that before I am at the end of this I shall have to repeat the same thing several times because long periods of time elapse between one writing and another [23, p. 41].

Leonardo never edited his manuscripts. Buried in one is his solution to the problem of the falling ball. It is in unconventional terms because modern

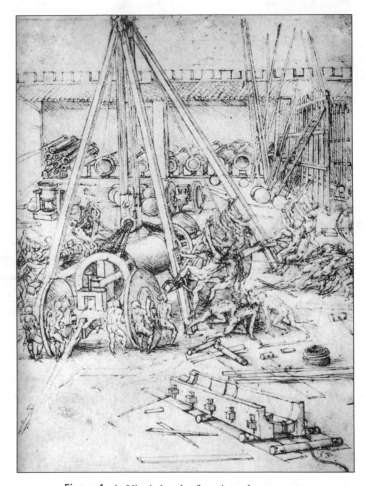

Figure 1. da Vinci sketch of casting a large cannon

conventions hadn't yet been developed. He asserted [**23**, p. 569–74]:

(i) The heavy thing descending freely gains a degree of speed with every stage of movement, although I ought first to say that with every degree of time it acquired a degree of movement more than the degree of the time immediately past.

(ii) If many bodies of similar weight and shape are allowed to fall one after the other at equal spaces of time the excesses of their intervals will be equal to each other.

In conventional terms:

(I) Let $v(t)$ be the speed at which an object is falling at time t. Then $v(t)$ is proportional to time, that is, $v(t) = kt$, where k is some constant.

(II) Let $s(t)$ be the distance an object falls in time t from rest. The second derivative of $s(t)$ is a constant.

Statement (I) is a clear translation of statement (i). In fact, Leonardo illustrates (i) by giving an arithmetic sequence of velocity values [23, p. 569]. A little work is required to see why statement (II) is a translation of statement (ii). Let us release weights at equally spaced time intervals: $t_0, t_1, \ldots, t_n$ so that $\Delta t = t_i - t_{i-1}$, $1 \le i \le n$. Interpret the word *interval* as the distance between successive falling objects. Immediately after dropping a new body, the intervals L_i are $L_i = s(t_i) - s(t_{i-1})$, $1 \le i \le n$. Interpret the *excess* of the intervals as the difference between successive intervals. Immediately after dropping a new body, the excesses $E(i)$ are $E_i = L_{i+1} - L_i = s(t_{i+1}) - 2s(t_i) + s(t_{i-1}) = c$, a constant, where $1 \le i < n$. Then

$$\frac{E_i}{(\Delta t)^2} = \frac{\dfrac{s(t_{i+1}) - s(t_i)}{\Delta t} - \dfrac{s(t_i) - s(t_{i-1})}{\Delta t}}{\Delta t} = \frac{c}{(\Delta t)^2}. \tag{1}$$

The middle expression of (1) is an approximation of the second derivative of $s(t)$. Thus, the second derivative of the falling distance is a constant. Not bad for 1500!

For a characterization of an earlier falling rule, the Merton rule of falling from 1335, see Exercise 4.

Galileo's Solution

One hundred years after da Vinci, Galileo reached the same conclusion, stating it much more clearly, that distance fallen is proportional to the square of the time. Galileo's object of choice for falling bodies is the cannonball. His most fabled experiment was dropping cannonballs from the Tower of Pisa to test the ancient claim that heavier objects fall faster than lighter objects. In his *Dialogue Concerning the Two Chief World Systems* he speaks of firing cannonballs both east and west. Which goes further? Do they tie? Would unequal measurements mean that the earth rotates? We'll leave this discussion for a later chapter.

Meanwhile, let's fire off a cannonball at time $t = 0$ on a flat Italian plain from position $A = (0, 0)$ at angle of elevation α and initial speed v_0.

Figure 2. Galileo's fabled experiment

We shall use conventional language to derive Galileo's solution, which he induced from experimentation. Let $x(t)$ and $y(t)$ be the x-coordinate and y-coordinate of the cannonball at time t. Let $\mathbf{s}(t) = (x(t), \ y(t))$ be the vector giving the position of the cannonball at time t. The velocity and acceleration of the cannonball at time t are respectively $\mathbf{v}(t) = \mathbf{s}'(t) = (x'(t), \ y'(t))$ and $\mathbf{a}(t) = \mathbf{v}'(t) = \mathbf{s}''(t) = (x''(t), \ y''(t))$. We assume that acceleration is a constant, $g = 32$ ft/sec^2 downwards. Thus, the initial conditions for the cannonball are

$$\mathbf{a}(t) = (0, -g) \text{ for all } t, \ \mathbf{v}(0) = v_0(\cos\alpha, \sin\alpha), \text{ and } \mathbf{s}(0) = (0, 0). \quad (2)$$

Integrating $\mathbf{a}(t)$ with respect to time and using the initial condition on $\mathbf{v}$ gives

$$\mathbf{v}(t) = (v_0 \cos\alpha, \ -gt + v_0 \sin\alpha). \quad (3)$$

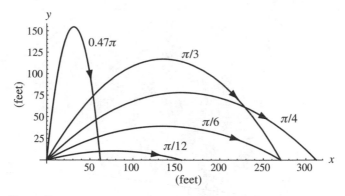

Figure 3. Trajectories, different angles, same speed: $v_0 = 100$ ft/sec

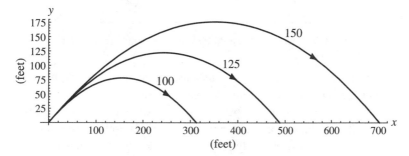

Figure 4. Trajectories, same angle: $\alpha = \pi/4$, different speeds

Integrating $\mathbf{v}(t)$ with respect to time and using the initial condition on $\mathbf{s}$ gives

$$\mathbf{s}(t) = \left(v_0 t \cos\alpha, \quad -\frac{1}{2}gt^2 + v_0 t \sin\alpha \right). \tag{4}$$

By (4), the time t_1 at which impact occurs is the non-zero solution of $-1/2\, g\, t^2 + v_0\, t\, \sin\alpha = 0$, $t_1 = 2v_0 \sin\alpha/g$. Therefore the range of the cannonball is $x(t_1)$ and its maximum height is $y(t_1/2)$. Figure 3 is a plot of various trajectories in which v_0 is constant and α varies. As can be seen, the range is a maximum when $\alpha = \pi/4$. Figure 4 is a plot of various trajectories for constant α. As v_0 increases, range and maximum height increase as well.

A Newtonian Solution

If a cannonball is fired with great velocity, we need a better model than Galileo's to account for the change in earth's gravitational acceleration as distance from the surface increases. Let us use Newton's inverse square law

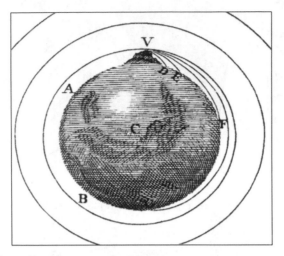

Figure 5. Newton's illustration of firing a cannon [**58**, p. 551]

that the acceleration of gravity at a distance r from the center of the earth is $a(r) = -gR^2/r^2$, where R is the radius of the earth.

Figure 5 is from Newton's *Treatise of the System of the World* showing cannonball trajectories for shots with varying initial velocities from atop a tall mountain. For a sufficiently large initial velocity, the cannonball never lands—it keeps rotating the earth! We will analyze trajectories involving spinning motion in the next chapter. Now we solve a simpler problem. We will shoot the cannonball straight up from a stationary earth and see what happens.

As in Chapter II,

$$v\frac{dv}{dt} = a\frac{ds}{dt}, \tag{5}$$

where s is distance and t is time, so

$$\int_0^\tau v\frac{dv}{dt}\,dt = \int_0^\tau a\frac{ds}{dt}\,dt. \tag{6}$$

At $t = 0$, assume that $s = R$ and that $v_0 > 0$. On the way up, s is monotonically increasing. Let $s(\tau) = r$. By the chain rule, (6) simplifies as

$$\frac{v^2(r) - v_0^2(R)}{2} = \int_R^r -\frac{gR^2}{s^2}\,ds = \frac{gR^2}{r} - gR,$$

which can be solved for v,

$$v = \sqrt{\frac{2gR^2}{r} - 2gR + v_0^2}. \tag{7}$$

If $v_0 < \sqrt{2gR}$, v reaches 0 and the cannonball then falls back to earth. Otherwise the projectile escapes the earth, which means that r increases without bound and so $\lim_{t\to\infty} v = \sqrt{v_0^2 - 2gR}$. The least possible value of this limit is 0. The least possible escape velocity from the earth is thus given by

$$v_0 = \sqrt{2gR}, \tag{8}$$

about 6.96 miles per second for the earth.

To find the time t at which the cannonball rises to distance r, use (18) of Chapter II. That is,

$$t = \int_R^r \frac{1}{v}\, ds, \tag{9}$$

where $v = \sqrt{2gR^2/s - 2gR + v_0^2}$ from (7). Despite the formidable-looking integral, t is an elementary function in terms of r. From (7) and (9) we can find v and t at any radial distance r, as long as v is positive.

A case study: Verne's moon shot

The greatest fictional ballistics shot of all time was described by Jules Verne. When he began his science fantasy writing, Americans were killing each other in the Civil War, and doing so with alarming inventiveness. The old muzzle loading rifle with a black powder-wadding-ball was modified with minie-ball technology, a harbinger of the cartridge. Ships sported armor. The first ironclads Monitor and Merrimac were launched. Cannon size and projectile range increased. Were there any limits to such weaponry?

Verne's 1865 *From the Earth to the Moon* began as a parody. In his story, the Civil War had just ended. American gun makers were idled. Gone was the need for more powerful cannonry. What were they to do? Of what use were they to society? Why not build a gargantuan 900 yard-long cannon and literally shoot the moon, sending a projectile reflective enough to be seen by telescopes? As Verne wrote, he got caught up in the excitement of his vision and redirected the story, having a Frenchmen—a Verne incarnation—convince the gunmakers to change the projectile into a capsule.

Figure 6. Verne's 1865 capsule, the Columbia

Verne tried to get his facts right in as far as his stories allowed, after which he let fiction fill in the details. A little more than 100 years after *From the Earth to the Moon* was published, America's Apollo spacecraft reached the moon. The similarities and contrasts between Verne's fiction and Apollo's facts are striking. Consider these points.

- Great fanfare accompany both announcements. In each case the world was riveted.

- Verne's capsule cost $5 million, while the total cost for the Apollo program was $20.5 billion. Using the relative gross domestic product indicator formula as given by [**97**], one dollar in 1865 has the equivalent worth of about $100 in 1968. Since the Apollo program involved about fifteen different launches, Verne's launch cost roughly a third of an Apollo launch.

- Florida was the launch site: Tampa in Verne's story, Cape Canaveral for Apollo.

- Preliminary launches involved animals. Verne sent up a cat and a squirrel, and only the cat returned, as it ate the squirrel during the flight. America and Russia sent up dogs and chimpanzees.

- Verne's three astronauts entered their craft using a crane; Apollo's three astronauts used an elevator. Great crowds watched both the launches. Verne's astronauts reclined on couches for the blast-off. Verne used a count-up to 40 (seconds) rather than a count-down.

- Verne's craft was 9 feet in diameter, cylindrical, with a conical nose, aluminum skin less than 2 inches thick. The Apollo craft was conical, almost 13 feet in diameter, 11 feet high.

- Flight to the moon took about four days for both Verne and Apollo. Apollo 8 orbited the moon in December 1968.

- Both spacecraft used retro-rockets for minor maneuvering.

- They both splashed down in the Pacific and were retrieved by ship.

- Verne's cannon was called the Columbiad. The command ship of Apollo 11 was called the Columbia.

In light of this last point, let's christen Verne's capsule as the Columbia, illustrated in Figure 6.

To determine the minimum velocity v_0 for Columbia to make moonfall, we simplify by assuming that the earth and the moon are fixed in space, saving the rotational version for the next chapter. Let r be the distance of Columbia from the earth's center. Let $D = 222,000$ miles be the distance between the earth's and moon's centers when the moon is at perigee, its nearest approach to the earth. Then $D - r$ is the distance of Columbia from the moon's center. The moon's mass is $\mu = 0.0123$ that of earth's. By Newton's second law of motion and the inverse square law,

$$m \frac{dv}{dt} = -\frac{GMm}{r^2} + \frac{\mu GMm}{(D-r)^2},$$ (10)

where G is the universal gravitation constant and M is earth's mass. Since $g = GM/R^2$ and since $\frac{dv}{dt} = \frac{dv}{dr}\frac{dr}{dt} = v\,v'$, (10) becomes

$$v\,v' = -gR^2 \left(\frac{1}{r^2} - \frac{\mu}{(D-r)^2} \right).$$ (11)

Integrating (11) and solving for v in terms of r gives

$$v(r) = \sqrt{2gR^2 \left(\frac{1}{r} + \frac{\mu}{D-r} \right) + v_0^2 - 2gR^2 \left(\frac{1}{R} + \frac{\mu}{D-R} \right)}.$$ (12)

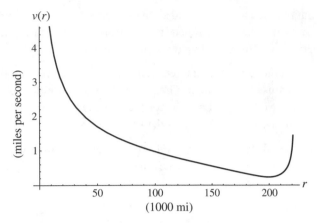

Figure 7. Velocity of Columbia, $v_0 = 6.89$ mi/sec

By (11), v is a minimum when

$$\frac{1}{r^2} - \frac{\mu}{(D-r)^2} = 0,$$

which means that $v_{\min}$ occurs when r is $r_\mu = D(1 + \sqrt{\mu}) \approx 199{,}837$ miles. Any initial velocity v_0 for which $v_{\min}$ exceeds zero will get Columbia to the moon.

Solving $v(r_\mu) = 0$ for v_0 gives 6.88605, or about 6.89 miles/sec, which is 123 yards/sec slower than escape velocity.

Figure 7 is a graph of velocity versus distance r. Let the radius of the moon be $L = 1082$ miles. The time T for Columbia to travel from the earth ($r = R$) to the moon ($r = D - L$) is

$$T = \int_R^{D-L} \frac{1}{v(r)} \, dr. \tag{13}$$

With $v_0 = 6.89$, $T \approx 3.84$ days. As v_0 approaches 6.88605, T increases without bound.

On their web site [57], NASA engineers analyze some of Verne's assumptions. They conclude that 180 linear yards of explosive cotton wadding in Columbiad's firing chamber would give only about 10% of the initial velocity needed to reach the moon, and they also say that the initial explosion would squash all life within the capsule.

Verne knew that some of what he proposed was pure fiction. Consider this dialogue from the story [90, p. 635]:

"But, unhappy man, the dreadful recoil will smash you to pieces at your starting."

"My dear contradictor, you have just put your finger upon the true and the only difficulty; nevertheless, I have too good an opinion of the industrial genius of the Americans not to believe that they will succeed in overcoming it."

Verne was right. However, it took a hundred years to fulfill his prophecy.

In the next chapter we will account for earth's rotation and modify the formulas of this chapter appropriately. As a case study, we will revisit the trajectory of Verne's Columbia as it goes from the earth to the moon.

Exercises

1. In Jules Verne's tale *Journey to the Center of the Earth*, three adventurers are blasted from a depth of 100 miles in the earth by a volcanic eruption while riding a raft of petrified logs. Imagine that the volcanic eruption functions as a cannon, giving the raft an initial velocity of v_0. In this case, the volcano's caldera is like the barrel of the cannon, so the raft behaves as a projectile. Assume that acceleration due to gravity is proportional to radial distance from the earth. Find v_0 so that the raft rises to the earth's surface, ending with zero velocity at sea level.

2. As Columbia is fired from Columbiad, imagine that she experiences a constant acceleration α within the 700 yards of the Columbiad's muzzle, so that her velocity is $v(t) = \alpha t$, where at $t = 0$ the 200 linear yards of explosive material is lit. Let T be the time at which Columbia exits the mouth of Columbiad. Find α and T given that $v(T) = 14{,}000$ yards/second. (In his book, Verne gives the launch velocity as 12,000 yards/second.) Then calculate the weight of a 180 pound astronaut undergoing such an acceleration.

3. In Verne's story, Columbia reaches the balance point between the moon and the earth, where their gravitational attractions are equal, in 83 hours and 20 minutes. Use this information to determine v_0.

4. In 1335, William Heytesbury, who taught at Merton College in Oxford, stated the Merton Rule of falling,

   The moving body, during some assigned period of time, will traverse a distance exactly equal to what it would traverse in an

equal period of time if it were moved uniformly at its mean degree
[**41**, p. 270].

Interpret the word *degree* as velocity. Imagine that the moving body in
the above passage falls from rest. Show that the Merton rule implies that
falling distance is proportional to the square of the time. This exercise
was adapted from [**36**].

5. Imagine being a muse to Leonardo da Vinci with respect to (1). Advise
   him by writing expressions for L, the distance interval between succes-
   sively falling objects, and E, the excess between successive distance
   intervals, as functions in terms of $t = i\Delta t$, the total time lapse from
   the beginning of the fall. Use the conventional notation $s'(t)$ and $s''(t)$.
   Assume that gravitational acceleration is $g = 32$ ft/sec^2.

6. From (3) and (4), explain how to deduce the initial velocity of a cannonball given by a specified powder charge in a particular cannon.

7. Consider this passage from Aristotle's *On the Heavens*:

   Heavy objects, if thrown forcibly upwards in a straight line, come
   back to their starting-place, even if the force hurls them to an
   unlimited distance [**5**, book ii, chapter xiv, p. 245].

   Show that Aristotle is correct about escape velocity

   (a) assuming that gravitational acceleration is a constant;
   (b) assuming that gravitational acceleration varies inversely with re-
       spect to distance from the earth's center.

8. Find the angle for which the arclength of a cannonball's trajectory is a
   maximum.

9. For Verne's moon shot with $v_0 = 6.89$ mi/sec, determine Columbia's
   speed when she reaches the moon. By how much does this velocity
   surpass the moon's escape velocity? Imagine being able to change the
   mass of the moon. Is there a mass size for the moon and an initial
   velocity v_0 from the earth's surface so that Columbia is captured by the
   moon and the moon trip is no more than four days?

10. Imagine two stars, A and Ω, whose centers are one billion miles apart.
    Suppose that A is a clone of our sun, and that Ω is smaller, with half the
    radius and half of the mass. Imagine that a huge solar flare explosion
    on Ω propels a mass P the size of our earth towards A. Determine the
    minimum initial velocity v_0 for P to reach A.

preamble iv

I'll turn over a new leaf.

Miguel de Cervantes (1547–1616),
Don Quixote

Early each Monday morning for a long season, my two preschool sons climbed down from their bunks, rushed to the living room couch, climbed its cushions and stood with bare feet, their hands over the back of the couch, peering out the picture window. The trash truck would soon appear! The trash men rode on the rear of the truck standing on a bumper, holding onto a rail with heavy leather gloves. As the truck slowed periodically with grinding of gears and mashing of its load, the men hopped off and tossed large cylindrical trash-filled containers over their heads, swinging them about as if they were stuffed animals. One day my boys wanted to be like these heroes.

For awhile I was concerned. Every father has great expectations for his children. Then I remembered my own dreams.

When I was about twelve, the space race between Russia and America had captured many imaginations. My dream, along with a horde of other youngsters, was to be an astronaut. I assembled plastic models of the Mercury, Gemini, and Apollo capsules, and suspended them from my bedroom ceiling. I clipped articles about NASA's space flights, pasting them into notebooks. I was numb along with the nation when Grissom, White, and Chafce were incinerated on a launch pad. I watched a news anchorman manipulate his own models demonstrating extra-vehicular activity and docking maneuvers.

For months I saved my allowance. When I had enough, I bought a telescope. With it I projected the sun onto a wall to see sunspots, and I used it to split double stars. My walls were decorated with posters of the solar system. I discovered science fiction, and read the masters: Verne, Wells, Asimov, Heinlein, Clarke.

Then I took my first plane ride.

I knew then that I was not astronaut material. Astronauts must endure rotation, speed, sudden jerks, and swerves. Not me.

No, I couldn't be an astronaut. But what about those behind the scenes, the ones who figured things out? The ones like Archimedes, Kepler, Galileo, Newton, Einstein. Could I be like them?

Every dream can turn in new directions.

CHAPTER IV

Heavenly Motion

Throw a ball. Where will it go? Now we take rotation into account.

We saw in the last chapter that in the early seventeenth century Galileo implicitly quantified acceleration due to gravity near the surface of the earth. Perhaps this was part of the elusive mechanism which kept planets in orbit? In the mid-seventeenth century, Christiaan Huygens quantified a new kind of acceleration, one that seemed to make heavenly motion more confusing. Shortly thereafter, Newton resolved the confusion with his publication of the *Principia*. In this chapter, we discuss these two developments, and then revisit Verne's moonshot of the previous chapter as a case study.

Huygens' acceleration

David killed Goliath with a small stone launched from a sling whirled over-head. Imagine an ideal version of David's sling: a pebble P rotating uniformly about the origin O at ω radians per second at fixed distance r from O. At time $t = 0$, release the pebble from its sling. Following Galileo's lead, Huygens believed he could find an expression for the acceleration for circular motion by examining the behavior of the pebble at the moment of release, at $t = 0$. Here is a conventional version of Huygens' argument, adapted from Groetsch [**36**].

Suppose that the distance of the pebble from its circle of rotation after it is released is given by $\hat{s}(t) = at^2/2$, for positive t near 0 for some expression a. We wish to find a at $t = 0$. From observation, the pebble flies off along a tangent to its circle of rotation, as illustrated in Figure 1.

Thus $\cos(\omega t) = r/(r + \hat{s}(t))$. That is,

$$\hat{s}(t) = r\,(\sec(\omega t) - 1) = \frac{at^2}{2}. \tag{1}$$

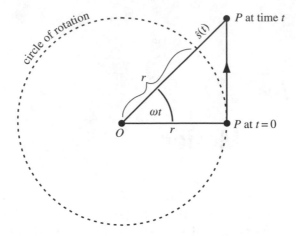

Figure 1. An ideal sling

Huygens was interested in this relationship only at $t = 0$. Solve (1) for a and take the limit as t goes to 0:

$$\lim_{t \to 0^+} a = \lim_{t \to 0^+} \frac{2r(\sec(\omega t) - 1)}{t^2}. \tag{2}$$

Two applications of L'Hôpital's rule show that the value of a at $t = 0$ is $\omega^2 r$.

To see that Huygens's argument gives the right answer, write the pebble's position about its circular orbit as

$$\mathbf{s}(t) = r(\cos(\omega t), \sin(\omega t)). \tag{3}$$

The pebble's acceleration $\mathbf{a}(t)$ is the second derivative of (3),

$$\mathbf{a}(t) = -r\omega^2(\cos(\omega t), \sin(\omega t)),$$

whose magnitude is indeed $a = \omega^2 r$. This value of a is called the centripetal acceleration of the pebble about its circular path.

Huygens' result posed a problem. The planets had fairly circular orbits, so they experienced this Huygensian acceleration away from the sun. Some other force must be counteracting the force arising from centripetal acceleration. But what?

In a dialogue about the moon, Plutarch flirted with these ideas long ago. Along with allusions to Atlas supporting the earth and lunar inhabitants tumbling off the moon due to its rotational velocity about earth, he had this fleeting foresight of both centripetal and gravitational forces:

All the same, you fear for the moon lest it fall [on] the Ethiopians. Yet the moon is saved from falling by its very motion and the rapidity of its revolution, just as missiles placed in slings are kept from falling by being whirled around in a circle. [Furthermore] the influence of [the moon's] weight alone might reasonably move the earth [**63**, p. 57–9].

Newton's Solution

Before the late seventeenth century, there was no mathematical explanation of how planets moved about the sun. Gravitational acceleration was thought to be a key. But what was gravity, and what causes it?

Could it be the masses of the objects? Could it be that all things attract one another? If so, it is reasonable that the greater the mass of a body, the greater is its gravitational attraction to other masses. The simplest assumption is that the variation is linear. Also, gravitational attraction should be radially symmetric, so the attraction between masses should depend only on the distance between them, and not their orientation in space.

Kepler had reasoned this way in the *Somnium*, an imaginative trip to the moon.

I define gravity as a power similar to magnetic power—a mutual attraction. The attractive force is greater in the case of two bodies that are near to each other than it is the case of bodies that are far apart [**44**, p. 106, footnote 66].

He goes on to describe, in footnote 67, that the power of gravity also depends on what we refer to as mass.

Let us construct a naïve model of gravity. Let P be a point-mass, and let B_r be a sphere of radius r with P at its center. Suppose P exudes an accelerative aura towards itself that is radially symmetric. Furthermore, suppose that this aura is conserved in the sense that the total attractiveness on the surface of B_r is the same for all r. Since the surface area of a ball of radius r is $4\pi r^2$, the totality of the aura at any patch of area A on the surface of B_r is $1/r^2$ that of any patch of area A on B_1, an inverse square relation! With the assumption of linearity in mass, we have Newton's gravitational acceleration function,

$$a(r) = \frac{km}{r^2}, \tag{4}$$

giving the acceleration which P induces on any point-mass at distance r from P, where k is a constant and m is the mass of P.

In 1684, with something like this in mind—for his research included investigation of magnetism, a phenomenon whose intensity also tails off according to the inverse square law—Edmond Halley asked Newton what orbit a body would follow if gravitational acceleration varies as the inverse square of distance. Newton replied that he had already calculated it, and that it was Kepler's ellipses, with the sun at a focus. Within three years, Newton's *Principia* appeared.

However, as is pointed out in Porciau [**65**, p. 7]:

> [The Principia] had a reputation in 1687; it has a reputation still—a reputation for being impenetrable.

Rather than giving Newton's string of geometrical propositions, we give the one that appears more or less in every calculus and mechanics text.

Position the polar plane with the sun's center at the origin and let the earth's position be (r, θ), where both r and θ are functions of time t. Let $\mathbf{s}$ be the rectangular coordinate equivalent to (r, θ). The two natural orthogonal unit vectors are $\mathbf{u}_r = \cos(\theta)\,\mathbf{i} + \sin(\theta)\,\mathbf{j}$ and $\mathbf{u}_\theta = -\sin(\theta)\,\mathbf{i} + \cos(\theta)\,\mathbf{j}$, as depicted in Figure 2.

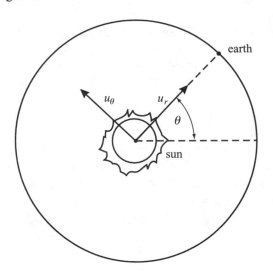

Figure 2. The natural directions

Since $\mathbf{s} = r\,\mathbf{u}_r$, $\frac{d\mathbf{u}_r}{d\theta} = \mathbf{u}_\theta$, and $\frac{d\mathbf{u}_\theta}{d\theta} = -\mathbf{u}_r$, earth's velocity is

$$\mathbf{v} = \frac{d\mathbf{s}}{dt} = \frac{d\,r\mathbf{u}_r}{dt} = r\mathbf{u}_\theta \frac{d\theta}{dt} + \mathbf{u}_r \frac{dr}{dt}.$$

Similarly, earth's acceleration is

$$\mathbf{a} = \left(r\frac{d^2\theta}{dt^2} + 2\frac{dr}{dt}\frac{d\theta}{dt} \right) \mathbf{u}_\theta + \left(\frac{d^2r}{dt^2} - r(\frac{d\theta}{dt})^2 \right) \mathbf{u}_r. \tag{5}$$

Assume that the earth's only acceleration is that induced by the gravitational field of the sun. Let $f(r) = -k/r^2$ be the gravitational acceleration r units from the sun, where k is some constant. Acceleration is solely in the radial direction, so

$$0 = r\frac{d^2\theta}{dt^2} + 2\frac{dr}{dt}\frac{d\theta}{dt} = \frac{1}{r}\frac{d\left(r^2\frac{d\theta}{dt}\right)}{dt} \quad \text{and} \quad -\frac{k}{r^2} = \frac{d^2r}{dt^2} - r\left(\frac{d\theta}{dt}\right)^2. \tag{6}$$

The first of the equations in (6) is the law of the conservation of angular momentum. Since $1/r$ is never 0,

$$r^2\frac{d\theta}{dt} = h, \tag{7}$$

where h is a constant. In terms of h, the second equation in (6) is

$$\boxed{\frac{d^2r}{dt^2} - \frac{h^2}{r^3} = -\frac{k}{r^2}.} \tag{8}$$

The general differential equation governing earth's motion

To solve (8) we need some initial conditions. Throughout its orbit, earth's distance from the sun varies, from a minimum value, the perigee, ρ, to a maximum value, the apogee. At time $t = 0$, we take $\theta = 0, r(0) = \rho$, which also means that $\frac{dr}{dt}|_{t=0} = 0 = \frac{dr}{d\theta}|_{\theta=0}$. The rate at which θ changes with respect to time is non-constant. It is a maximum at perigee and a minimum at apogee. Since θ changes by 2π radians over a year, we let $\frac{d\theta}{dt}|_{t=0} = \omega_0$, where $\omega_0 > 2\pi$ radians/yr. To summarize,

$$\boxed{\begin{array}{l} r = \rho \text{ when } t = 0 \text{ and } \theta = 0, \quad \text{and} \\ \frac{dr}{dt}\Big|_{t=0} = 0 = \frac{dr}{d\theta}\Big|_{\theta=0}, \text{ and } \frac{d\theta}{dt}\Big|_{t=0} = \omega_0. \end{array}} \tag{9}$$

The initial conditions

A clever way to solve (4) for r in terms of θ is to let $z = 1/r$. Then, by (7),

$$\frac{dr}{dt} = \frac{d(\frac{1}{z})}{d\theta}\frac{d\theta}{dt} = -\frac{1}{z^2}\frac{dz}{d\theta}\frac{d\theta}{dt} = -r^2\frac{d\theta}{dt}\frac{dz}{d\theta} = -h\frac{dz}{d\theta}.$$

Similarly,

$$\frac{d^2r}{dt^2} = -h^2z^2\frac{d^2z}{d\theta^2}.$$

Thus another way to write (4) is

$$\frac{d^2z}{d\theta^2} + z = \frac{k}{h^2}. \tag{10}$$

The homogeneous analog to (10), $\frac{d^2z}{d\theta^2} + z = 0$, is the celebrated differential equation governing simple harmonic motion. For the reader unfamiliar with solving (10), see the simple harmonic motion entry in the appendix. The solution to the homogeneous equation is $z_0 = A\cos(\theta + \phi)$, where A and ϕ are constants. An obvious particular solution to (10) is $z_1 = k/h^2$. Thus the general solution to (10) is $z = z_0 + z_1 = A\cos(\theta + \phi) + k/h^2$, so the general solution to (8) is

$$\frac{1}{r} = A\cos(\theta + \phi) + \frac{k}{h^2}. \tag{11}$$

Differentiating (11) with respect to θ gives $1/r^2\frac{dr}{d\theta} = A\sin(\theta + \phi)$, which by (9) means that $A\sin\phi = 0$, so we can take the phase angle, ϕ, to be 0. Using (9) and (11) gives $1/\rho = A\cos(0) + k/h^2$, so $A = 1/\rho - k/h^2$. Thus the orbit of the earth is given by

$$r = \frac{h^2/k}{1 + \left(\frac{h^2}{k\rho} - 1\right)\cos\theta}. \tag{12}$$

Earth's motion about the sun

Provided $0 < h^2/(k\rho) < 2$, this is the equation of an ellipse.
The area $A(t)$ of this ellipse swept out in time from 0 to t is, from (7),

$$A(t) = \int_{\theta(0)}^{\theta(t)} \frac{1}{2}r^2\,d\theta = \int_0^t \frac{1}{2}r^2\frac{d\theta}{d\tau}\,d\tau = \int_0^t \frac{1}{2}h\,d\tau = \frac{ht}{2}. \tag{13}$$

This is Kepler's second law of motion.

The standard form of an ellipse in polar coordinates is

$$r = \frac{b^2/a}{1 + e\cos\theta}, \tag{14}$$

where a is the length of the semi-major axis, b is the length of the semi-minor axis, and e is the eccentricity $e = c/a$, where $c = \sqrt{a^2 - b^2}$. Its area is πab.

Let T be the period of the orbiting body. From (13), the area of the ellipse is $hT/2$. Comparing (12) and (14), we see that

$$b^2/a = h^2/k, \tag{15}$$

so $b/h = \sqrt{a/k}$. Thus

$$T = \frac{2\pi}{h}ab = 2\pi a\sqrt{\frac{a}{k}}.$$

This is Kepler's third law of motion,

$$T^2 = \frac{4\pi^2}{k}a^3. \tag{16}$$

From (15), $h = b\sqrt{k/a}$, and a, b, and k can be determined by astronomical observation.

To recap, (12), (13), and (16) in terms of any planet about its sun are

$$\text{Kepler's three laws where } f(r) = -\frac{k}{r^2}: \tag{17}$$

  i. Planet P's orbit is an ellipse with one focus at the sun's center.
 ii. P's position vector (with respect to the sun) sweeps out area at a constant rate.
iii. The square of P's period is proportional to the cube of the length of the semi-major axis.

Case study: Verne's Moon Shot revisited

Figure 3 is an engraving from the 1865 edition of Jules Verne's *From the Earth to the Moon,* showing Columbia's blast-off, that was initiated by a spark into the breech of the cannon.

An appalling, unearthly report followed instantly, such as can be compared to nothing whatever known, not even to the roar of thunder, or the blast of volcanic explosions! An immense spout of fire shot up from the bowels of the earth as from a crater. The earth heaved up, and with great difficulty some few spectators obtained a momentary glimpse of the projectile victoriously cleaving the air in the midst of the fiery vapors!

At the moment when that pyramid of fire rose to a prodigious height into the air, the glare of the flame lit up the whole of Florida; and for a moment day superseded night over a considerable extent of the country. This immense canopy of fire was perceived at a distance of one hundred miles out at sea [90, pp. 663–4].

As for the astronauts inside Columbia, they all lost consciousness for some minutes. The Frenchman awoke first and "tried to rise, but could not stand. His head swam, from the rush of blood; he was blind; [the blast-off produced] the same effect as swallowing two bottles of Corton" [90, p. 682].

In Chapter III we analyzed Columbia's trajectory while ignoring the rotation of both the earth and the moon. Now we take them into account.

For simplicity, we will assume that the moon orbits the earth in a circle whose center is the center of the earth. We ignore the sun altogether and assume that the earth's center remains fixed in space. Assume that the distance from the earth's center to the moon's center is ρ, and that the moon rotates uniformly about the earth once every $Q = 27.3$ days. Let $\omega = \omega(t) = (2\pi/Q)(t - t_0)$, where t_0 is some constant. The moon's position $M(t)$ is thus

$$M(t) = \rho(\cos\omega(t), \sin\omega(t)). \tag{18}$$

Columbia's position at time t is

$$P(t) = r(t)(\cos\theta(t), \sin\theta(t)). \tag{19}$$

Let $\mu = 0.0123$, the relative mass of the moon with respect to the earth. Let L be the distance of Columbia from the center of the moon. That is, $L^2(t) = (\rho\cos\omega(t) - r\cos\theta)^2 + (\rho\sin\omega(t) - r\sin\theta)^2$. The gravitational attraction on Columbia as induced by the moon has magnitude $\mu gm R^2/L^2$, where $R = 4000$ miles is the radius of the earth and m is the mass of Columbia. The unit direction of this force is $(M(t) - P(t))/L(t)$. (See the entry on projection in the appendix.) The component of the moon force in the $\mathbf{u}_\theta$ direction is

$$m\,A_{M,\theta} = \left(\frac{\mu gm R^2}{L^2}\right)\left(\frac{M(t) - P(t)}{L}\right)\cdot\mathbf{u}_\theta$$

$$= \frac{\mu g\rho m R^2}{L^3}(-\cos\omega\sin\theta + \sin\omega\cos\theta),$$

Figure 3. Columbia at launch

and in the $\mathbf{u}_r$ direction is

$$m\,A_{M,r} = \left(\frac{\mu g m R^2}{L^2}\right)\left(\frac{M(t) - P(t)}{L}\right) \cdot \mathbf{u}_r$$

$$= \frac{\mu g m R^2}{L^3}(\rho\cos\omega\cos\theta + \rho\sin\omega\sin\theta - r).$$

The corresponding components of the earth force on Columbia are

$$m\,A_{E,\theta} = 0 \qquad \text{and} \qquad m\,A_{E,r} = -\frac{g m R^2}{r^2}.$$

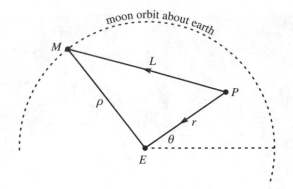

Figure 4. Columbia pulled by both the earth and the moon

Therefore the motion of Columbia satisfies the system of differential equations

$$\begin{cases} r\theta'' + 2r'\theta' = A_{M,\theta}, \\ r'' - r(\theta')^2 = A_{E,r} + A_{M,r}, \end{cases} \tag{20}$$

where the derivatives are with respect to t. The equations are independent of m.

Since we are unable to solve (20) analytically, we will use Euler's method for solving systems of differential equations as outlined in Chapter I. This time the moon is a moving target. We will launch Columbia at time $t = 0$,

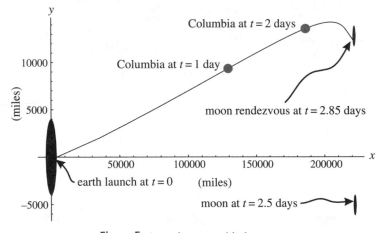

Figure 5. A rendezvous with the moon

with $R = r(0) = 4000$ miles, $\theta(0) = 0$, $\frac{dr}{dt}|_{t=0} = 6.9$ miles per second, $\frac{d\theta}{dt}|_{t=0} = 2\pi/T$, where T is one day, $\rho = 222{,}000$ miles, and $t_0 = 2.6$, which means that when Columbia is launched, the moon is 2.6 days from reaching $\theta = 0$. Besides its initial blast-off velocity of 6.9 mi/sec, Columbia has a tangential velocity of $2\pi R/T \approx 0.3$ mi/sec as given by earth's rotation, assuming that the launch point is at the equator rather than in Florida.

With all the initial conditions, Columbia and the moon rendezvous occurs in about 2.85 days, which is about a day faster than Verne predicted. Figure XIV shows this trajectory, not drawn to scale. The moon is the narrow ellipse at the trajectory's end point. Drawn to scale, the trajectory would be indistinguishable from the x-axis. In Verne's story, Columbia shoots on by the moon, missing its north pole by a mere four miles, which is a remarkably lucky shot considering that even with a computer it took some hunting to determine that t_0 should be about 2.6 days in order to have a moon rendezvous.

In the exercises that follow we suggest ways to improve the accuracy of the model.

Verne's book ends with a series of questions which are as valid as when he wrote them.

Will this attempt [circling the moon], unprecedented in the annals of travels, lead to any practical result? Will direct communication with the moon ever be established? Will they go from one planet to another, from Jupiter to Mercury, and after awhile from one star to another, from the Polar to Sirius? [**90**, p. 822]

The next few generations may answer them.

Exercises

1. Let e be a number with $0 < e < 1$. Show that

$$\int \frac{h^2/k}{(1 + e\cos\theta)^2}\, d\theta = \frac{2h^2/k}{(1-e^2)^{\frac{3}{2}}} \tan^{-1}\left(\sqrt{\frac{1-e}{1+e}} \tan\frac{\theta}{2}\right)$$
$$- \frac{eh^2/k\sin\theta}{(1-e^2)(1+e\cos\theta)}.$$

Use this, along with equations (7) and (12), to find t in terms of θ as the earth goes around the sun. Suppose at noon on the first of January, $\theta = 0$ and the earth is at perihelion. Find θ at noon on the first of April. Assume that the earth's period is exactly 365 days. *Hint:* To integrate, use the substitution $\theta = 2\tan^{-1} z$ so that $\cos\theta = (1 - z^2)/(1 + z^2)$ and $d\theta = 2/(1 + z^2)\, dz$.

2. Imagine that planet X has earth's orbit of 365 days about the sun, but with eccentricity $e = 0.5$. Assume that the winter solstice is on January 1. When is spring? That is, when is the spring equinox? Use the result of exercise 1.

3. In our version of Verne's moon-shot, we launched the Columbia at the equator. Now launch the Columbia as Verne did at Tampa, latitude 28°. How do the initial conditions change? Assume that motion remains in a plane. Determine the length of the moon trip for blast-off velocity of 5.9 mi/sec and compare the length to the 2.85 days of the text.

4. Verne's Columbia reaches the moon in 97 hours and 13 minutes. Find the initial blast-off velocity needed to achieve this time. Compare with Verne's initial blast-off velocity of 12,000 yards per second.

5. Assume that the moon orbits the earth about the earth's center in 27.3 days. Adapt the results of Exercise 1 to the moon-earth system. What is θ at noon on January 10?

6. For the results of Exercise 5, find a function which is a linear combination of 1, t, and $\sin(2\pi/27.3)$ that approximates θ for one period of the moon.

7. Use the conditions and results of Exercises 5 and 6 to re-solve Verne's moon shot problem using an elliptical orbit for the moon instead of a circular one.

8. Design a better moon-earth-Columbia system of differential equations model by having the earth and moon rotate about their common center of gravity. Then re-solve Verne's moon-shot problem.

9. Plot Columbia's trajectory using a sun-earth-moon system in which the earth rotates in a circle about the sun and the moon rotates in a circle about the earth. Compare your results with the example in the text for the earth-moon system. How much of a factor is the sun in this new model?

10. Repeat Exercise 10 of Chapter III assuming that the stars rotate about their common center of mass in one day.

preamble v

And all the way, to guide their chime,
With falling oars they kept the time.

Andrew Marvell (1621–1678), *Bermudas*

In the summer of 2004, a friend of twenty some years, a professor of English, had taken it on himself to remodel the inside of a house on campus that had suffered years of neglect. His idea was to transform it into an honors house, where students could have study carrels, hang out, have seminars, and discuss ideas.

Inside the house I looked at the tired paint, the broken windows, the corroded plumbing, and the disfigured floors, and I asked, "What about the outside?"

"Oh that," Craig laughed apologetically, "that will have to wait until next year."

Before I had gone inside, I had circled the house. The exterior paint had peeled. The clapboards were rotted or badly warped. Nails protruded haphazardly all around, having worked their way outwards over the years as

the building weathered in the rain and snow and Tennessee sun. The roof, however, was in good repair.

I knew Craig wasn't joking. I also knew that he had been appointed director of the honors program and he alone was responsible for getting the house in order.

"This won't do," I told him, "the outside must be done as well. An honors house can't look like a dump. I can do it in a month."

He accepted my offer.

I started at the top, working downwards, taking one side of the house each week. By the fourth week I was on the last side, the front. My ladder couldn't span the porch, so I built a platform with triangular cross-section to put on the slanted porch roof to give a solid horizontal surface for a step ladder, as shown on the previous page. But I hadn't noticed a particular mossy region of the roof. When I was standing near the top eaves, the platform slipped and the top of the ladder slid down the side of the house. I remember that as I fell, I observed with as much satisfaction as possible under the circumstances that the top of the ladder hit the roof away from the side of the house, just like the straight stick pendulum ladder of the next chapter. The ladder crashed over the eaves. I was left face down, spread-eagled on the shingles. Those who visit the honors house now, should they look closely at the porch roof, can see faint stains where the paint bucket landed.

CHAPTER V

Pendulum Variations

When part of a mechanical contrivance is allowed to fall about a fixed point, a pendulum is born. Whether it be a swing on the playground, a swaying metal rod within a grandfather clock, a balance scale in a science lab, they all try to teeter back and forth in a peculiar way.

In 1582, Galileo noted that a swinging lamp's period is the same regardless of its amplitude, so a pendulum could be used to tell time. However, none was built until 1657 when Christiaan Huygens made one that was accurate to within a minute per day. The Dutch stamp of Figure 1 commemorates this achievement. In 1667, Edmond Halley, while observing Mercury's transit from the top of a mountain, was the first to notice that a pendulum's period

Figure 1. The first pendulum clock, 1657

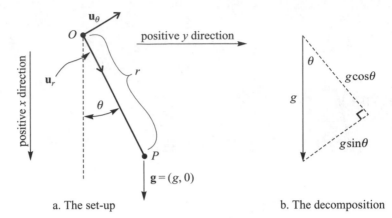

a. The set-up   b. The decomposition

Figure 2. A pendulum model

varies with altitude, which could only mean that gravitational acceleration varies, a prosaic observation today, but a momentous one in Halley's day.

In this chapter we shall see that the differential equation governing a pendulum's motion is the same as that governing how planets move in their orbits! We also consider the classic sliding ladder problem and find its equations of motion by looking at it as a pendulum.

The standard pendulum

The model is laid out in Figure 2a. Take a line segment of length r. Fix an endpoint, called the pivot, at the origin, O. Assume that all of the mass of the pendulum is at the other endpoint, P. In polar coordinates, $P(t) = (r, \theta)$ at time t where r and θ are functions of t and $\theta = 0$ is straight down. Let $\mathbf{u}_r = (\cos\theta, \sin\theta)$ and $\mathbf{u}_\theta = (-\sin\theta, \cos\theta)$. So, in rectangular coordinates with the positive x direction straight down, and the positive y direction to the right, $P(t) = r\,\mathbf{u}_r$. Then just as in (5) of Chapter IV,

$$P''(t) = (r'' - r(\theta')^2)\mathbf{u}_r + (2r'\theta' + r\theta'')\mathbf{u}_\theta. \tag{1}$$

Since r is a constant,

$$P''(t) = (-r(\theta')^2)\,\mathbf{u}_r + (r\theta'')\,\mathbf{u}_\theta. \tag{2}$$

To obtain a differential equation modeling the pendulum's motion, we decompose the gravitational acceleration g of the pendulum's mass into components in the $\mathbf{u}_r$ and $\mathbf{u}_\theta$ directions, writing $(g, 0)$ as $\alpha\mathbf{u}_r + \beta\mathbf{u}_\theta$ for some numbers α and β. Since $\mathbf{u}_r$ and $\mathbf{u}_\theta$ are orthonormal vectors, $\alpha = (g, 0) \cdot \mathbf{u}_r$

and $\beta = (g, 0) \cdot \mathbf{u}_\theta$. Therefore

$$(g, \ 0) = g \cos \theta \, \mathbf{u}_r - g \sin \theta \, \mathbf{u}_\theta. \tag{3}$$

Another way to find the decomposition in (3) is to determine the leg-lengths of the right triangle whose hypotenuse g is aligned with the vertical, and whose side adjacent to the angle θ lies in the direction of line $\overline{OP}$, as in Figure 2b.

The $\mathbf{u}_r$ component of (3) governs the internal tension of the line segment, a phenomenon that will not affect the pendulum's motion, and so can be ignored. Therefore, equating the $\mathbf{u}_\theta$ components of (2) and (3) gives the differential equation describing the motion of the pendulum:

$$r \, \theta'' = -g \sin \theta. \tag{4}$$

This differential equation cannot be solved analytically. However, if θ is small, never exceeding about $10°$, then a good approximation for $\sin \theta$ is θ, so an approximate model for the pendulum's motion is the differential equation

$$r \, \theta'' = -g \, \theta. \tag{5}$$

Its solution is

$$\theta(t) = A \cos(t\sqrt{g/r} + \phi), \tag{6}$$

where A and ϕ are constants determined by initial conditions, which is simple harmonic motion. (The simple harmonic motion entry in the appendix gives an outline for solving this equation.) The period T of the pendulum is

$$T = 2\pi \sqrt{r/g}, \tag{7}$$

a constant as long as one stays at the same altitude, as Galileo observed, and varies with g as gravity changes, as Halley observed.

In the remainder of this chapter we modify the model above to study two variations of the standard pendulum, a stick pendulum and a descending pendulum.

Case study: a sliding ladder

Consider the classic sliding ladder problem:

> A ladder r feet long leans against a vertical wall. If the base of the ladder is moved outwards at the constant rate of k feet per second, how fast is the tip of the ladder moving downwards?

The usual solution assumes that the tip of the ladder slips downward, maintaining contact with the wall until impact at ground level, so that the base and tip of the ladder at time t have coordinates $(x(t),\ 0)$ and $(0,\ y(t))$, respectively, as in Figure 3. Differentiating the relation $x^2 + y^2 = r^2$ given by the Pythagorean theorem, we find

$$y' = \frac{-kx}{y}. \tag{8}$$

A defect of this model is that as the ladder nears the ground, y' becomes so large that the speed of the tip of the ladder exceeds the speed of light (see Strang [**85**, p. 164]).

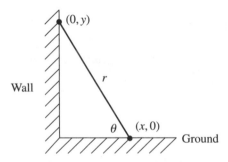

Figure 3. A sliding ladder model

Actually, the ladder's tip leaves the wall at some point in its descent.

We will determine y_c, the critical height at which the ladder leaves the wall and for which (8) ceases to be valid. We do this by examining the differential equations governing two different physical situations: the moving ladder supported by the wall and the unsupported ladder behaving as a stick pendulum.

For the pendulum, the rotational version of Newton's second law of motion states that if a rigid body rotates in a plane about an axis that moves with uniform velocity, then the total torque exerted by all the external forces on the body equals the product of the moment of inertia and the angular acceleration, where the torque and the moment of inertia are computed with respect to this moving axis. Torque is the product of force and the perpendicular distance from the line of action of the force to the axis of rotation. We define moment of inertia and angular acceleration in context in the next few paragraphs.

Since the pivot moves with constant velocity, we apply Newton's second law to the axis that passes through the point of contact of the ladder with the ground and that is perpendicular to the plane in which the ladder falls.

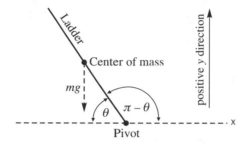

Figure 4. A straight stick pendulum of length r

The only forces on the freely falling ladder are the upward force from the ground at the pivot point, which produces no torque, and the gravitational force, which produces the same torque τ as a force of magnitude mg acting downward at the center of mass of the ladder, as indicated in Figure 4. That is,

$$\tau = mg\frac{r\cos\theta}{2},$$

a positive value since this torque is counterclockwise. The moment of inertia I about a pivot of a set of n point masses m_i which are x_i units from the pivot, $1 \le i \le n$, is

$$I = \sum_{i=1}^{n} x_i^2\, m_i.$$

Therefore the moment of inertia I of a uniform rod with mass m and length r about an endpoint is

$$I = \lim_{n\to\infty} \sum_{i=1}^{n} x_i^2 \left(\frac{m}{r}\,\Delta x\right) = \int_0^r x^2 \frac{m}{r}\,dx = \frac{1}{3}mr^2,$$

where $\Delta x = r/n$ and $x_i = i\,\Delta x$. The angular acceleration is $-\theta''$, the second derivative with respect to time of the angle $\pi - \theta$ between the ground and the ladder, measured counterclockwise. Thus, Newton's law for the falling ladder gives $\frac{1}{3}mr^2(-\theta'') = \frac{1}{2}mgr\cos\theta$, or

$$\theta'' = -\frac{3g}{2r}\cos\theta, \tag{9}$$

which is valid after the ladder loses contact with the wall.

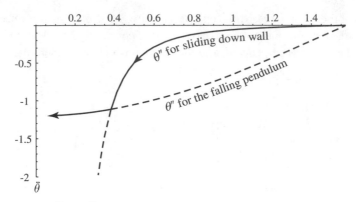

Figure 5. The transition between sliding and swinging

On the other hand, when the ladder is in contact with the wall, $y = r \sin \theta$, and differentiation yields $y' = r \cos(\theta) \, \theta' = x \, \theta'$. By (8)

$$\theta' = -\frac{k}{r \sin \theta}, \tag{10}$$

which yields

$$\theta'' = \frac{k \cos \theta}{r \sin^2 \theta} \theta' = -\frac{k^2 \cos \theta}{r^2 \sin^3 \theta}, \tag{11}$$

which is valid while the ladder maintains contact with the wall.

Given specific values of r and k, we can determine the critical angle θ_c at which the ladder loses contact with the wall by finding the point of intersection of the graphs of (9) and (11), plotting θ'' versus θ. Figure 5 illustrates this using the values $r = 40$ ft, $k = 10$ ft/sec, $g = 32$ ft/sec^2. From the graph we see that as θ decreases the ladder falls according to (11) until the two curves meet at $\theta_c \approx 0.38$, the critical angle, and thereafter the ladder falls according to (9). Up until the critical angle the ladder is held up by the wall, but after θ_c it is free to behave as a stick pendulum.

Leaving r and k as parameters, we equate the right sides of (9) and (11), and simplify to get

$$\sin^3 \theta_c = \frac{2k^2}{3gr}. \tag{12}$$

If $2k^2/(3gr) > 1$, that is, if $k > \sqrt{3gr/2}$, (12) is invalid, and we conclude that the tip of the ladder pulls away from the wall immediately once the base moves away with speed k. Otherwise, since $y_c = r \sin \theta_c$,

$$y_c = \sqrt[3]{\frac{2k^2 r^2}{3g}}. \tag{13}$$

It is interesting to find y_c'', the acceleration of the tip of the ladder at the critical height. Differentiating (8) and simplifying yields $y'' = -k^2 r^2/y^3$, which is valid while the ladder stays in contact with the wall. By (13) then, the acceleration at the moment of separation is

$$y_c'' = -\frac{3}{2} g. \tag{14}$$

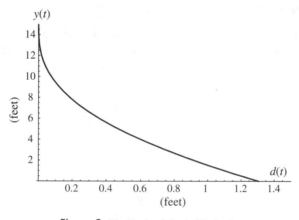

Figure 6. The path of the ladder's tip

To find the path of the ladder's tip after it leaves the wall, first observe that at the moment of separation the base is at $x_c = r \cos \theta_c$. Since the base moves away at constant speed k, its distance from the wall t seconds later will be $x_c + kt$, so the distance from the wall to the upper end of the ladder at this time will be $d = x_c + kt - r \cos \theta$. Thus, the path of the ladder's tip will be given by

$$\begin{cases} d(t) = x_c + kt - r \cos \theta(t), \\ \quad y(t) = r \sin \theta(t). \end{cases} \tag{15}$$

Figure 6 shows the trajectory as given by (9) where $r = 40$ ft, $k = 10$ ft/sec and $g = 32$ ft/sec^2. The initial values are

$$\theta(0) = \theta_c = \arcsin \sqrt[3]{\frac{2k^2}{3gL}} \approx 0.3827, \text{ from (12), and}$$

$$\theta'(0) = \dot{\theta}_c = -\frac{k}{r \sin \theta_c} \approx -0.6694, \text{ from (10).}$$

In this example $y_c = r \sin \theta_c \approx 14.94$ ft. The solution is computed as long as $\theta(t) \geq 0$, which turns out to be about 0.42 seconds, and at the moment of impact the distance of the tip of the ladder from the wall is $d \approx 1.30$ ft.

See Freeman [34] for a more traditional approach to modeling the sliding ladder.

Case study: Poe's pendulum

Perhaps the most macabre literary story featuring a pendulum is Edgar Allan Poe's "The Pit and the Pendulum." Its narrator is strapped to a table in a dungeon during the Inquisition. Suspended from a pivot about 35 feet overhead, a descending blade,

> a crescent of glittering steel, about a foot in length from horn to horn the under edge as keen as a razor, and the whole hissed as it swung through the air,

inches ever closer to the narrator [64, p.252]. In Poe's description, the pendulum's velocity and amplitude increase with descent. In particular, the pendulum increases from being "brief" and "slow" to "a terrifically wide sweep (some thirty feet or more)," that is, to an amplitude of about 15 feet. In an hour's time from first noticing its descent, "the sweep of the pendulum had increased by nearly a yard. As a natural consequence its velocity was also much greater."

To model Poe's pendulum, we follow Borelli [12] and assume that all of the mass of the pendulum is at the blade tip, P, and neglect resistance. With the model of Figure 2 and the same setup as in the standard pendulum at the beginning of this chapter,

$$P''(t) = (r'' - r(\theta')^2)\mathbf{u}_r + (2r'\theta' + r\theta'')\mathbf{u}_\theta.$$

In this case r varies with time and so the analog of (4) for the descending pendulum is

$$r\theta'' + 2r'\theta' + g\sin\theta = 0. \tag{16}$$

As Poe describes it, when the pendulum has length $L = 32$ feet, a length just above the narrator's torso, then the pendulum descends three inches during 11 periods of the pendulum.

> Down—certainly, relentlessly down! It vibrated within three inches of my bosom! I struggled violently—furiously—to free my left arm.

> I saw that some ten or twelve vibrations would bring the steel in actual contact with my robe—and with this observation there suddenly came over my spirit all the keen, collected calmness of despair [**64**, p. 254].

Figure 7. A scimitar of the Inquisition

Since the period T of a pendulum of length L is $T = 2\pi \sqrt{r/g} = 2\pi$ seconds by (7), the pendulum descends at the rate of about $\alpha = 1/(88\pi)$ ft/s. If the rate is constant, the time for the pendulum to descend from $r = 2$ to $r = 32$ is about 2.5 hours, consistent with the story, for Poe's narrator describes the total torturous descent as lasting several hours:

> What boots it to tell of the long, long hours of horror more than mortal, during which I counted the rushing oscillations of the steel! Inch by

inch—line by line—with a descent only appreciable at intervals that seemed ages—down and still down it came! [**64**, p. 253]

Thus (16) becomes

$$\alpha t\, \theta'' + 2\alpha\, \theta' + g \sin\theta = 0. \tag{17}$$

For initial conditions, let's say that when $r = 2$ feet, $t = 176\pi$ seconds, $\theta = 0$, and $\theta' = 1$ radian per second.

Numerically solving (17) as the pendulum's length changes from 2 to 32, and then parametrically plotting $r(t)(\cos\theta,\ \sin\theta)$ gives Figure 8. (In our model, remember that the x component is down and the y component is to the right.) As can be seen, an initial amplitude of about half of a foot at pendulum length two feet grows to about an amplitude of one foot at pendulum length 32 feet, nowhere near Poe's estimate of 15 feet. Furthermore, a similar plot of the pendulum's speed, which we leave for the reader, demonstrates that the speed of the blade is slowly decreasing rather than increasing according to Poe's description.

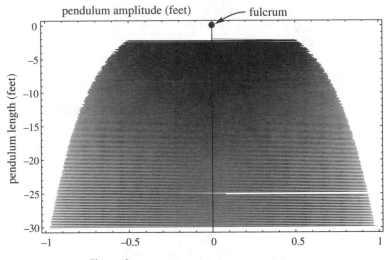

Figure 8. The amplitude of Poe's pendulum

If we use the convention that θ is always small, then a good approximation to (17) is

$$\alpha t\, \theta'' + 2\alpha\, \theta' + g\theta = 0. \tag{18}$$

However, from Figure 8, when $r \approx 2$, the pendulum's maximum amplitude is about 0.5 feet, which corresponds with $\theta \approx 14°$, which is a borderline for a small angle.

Nevertheless, let's try an analytic approach and transform (18) into a Bessel differential equation,

$$x^2 y'' + xy' + (x^2 - p^2)y = 0, \tag{19}$$

where p is a nonnegative real number. One transformation that works is $y = \sqrt{t}\,\theta$ and $x = k\sqrt{t}$, where $k^2 = 4g/\alpha$. Then

$$\frac{dx}{dt} = \frac{k}{2\sqrt{t}} \quad \text{and} \quad \frac{dy}{dt} = \frac{\theta}{2\sqrt{t}} + \sqrt{t}\,\theta',$$

so

$$z = y' = \frac{dy}{dx} = \frac{\left(\frac{dy}{dt}\right)}{\left(\frac{dx}{dt}\right)} = \frac{1}{k}\theta + \frac{2t}{k}\theta'. \tag{20}$$

Thus

$$\frac{dz}{dt} = \frac{3}{k}\theta' + \frac{2}{k}t\,\theta'',$$

so

$$y'' = \frac{d^2 y}{dx^2} = \frac{dz}{dx} = \frac{\left(\frac{dz}{dt}\right)}{\left(\frac{dx}{dt}\right)} = \frac{6}{k^2}t^{\frac{1}{2}}\theta' + \frac{4}{k^2}t^{\frac{3}{2}}\theta''. \tag{21}$$

Thus the Bessel equation $x^2 y'' + xy' + (x^2 - 1)y = 0$ becomes, from (20) and (21),

$$k^2 t\left(\frac{6}{k^2}t^{\frac{1}{2}}\theta' + \frac{4}{k^2}t^{\frac{3}{2}}\theta''\right) + kt^{\frac{1}{2}}\left(\frac{1}{k}\theta + \frac{2t}{k}\theta'\right) + (k^2 t - 1)t^{\frac{1}{2}}\theta = 0,$$

which simplifies to $t^2\theta'' + 2t\theta' + (k^2/4)t\theta = 0$, the original differential equation (18). Let $y = A\,J_1(x) + B\,Y_1(x)$ be the solution to (19) with $p = 1$, where $J_1(x)$ and $Y_1(x)$ are the standard Bessel functions of the first and second kind, and A and B are constants. See the Bessel function entry in the appendix for a more detailed representation of these two functions. When $t = 176\pi$, $\theta = 0$, and $\theta' = 1$, so $x = 1408\pi$, $y = 0$, and $y' = 0/k + (2/k)(176\pi)(1) = 352\pi/k = \sqrt{11\pi}$. If we use a computer algebra system, we find $A \approx 346.521$ and $B \approx -346.462$. Thus the solution to (18) is

$$\sqrt{t}\,\theta \approx A\,J_1(k\sqrt{t}) + B\,Y_1(k\sqrt{t}),$$

which we rewrite as

$$\theta(t) \approx \frac{346.521}{\sqrt{t}}\,J_1(32\sqrt{11\pi t}) - \frac{346.462}{\sqrt{t}}\,Y_1(32\sqrt{11\pi t}). \qquad (22)$$

The graph of $r(t)(\cos\theta, \sin\theta)$ for (22) is almost identical to Figure 8, demonstrating that the replacement of $\sin\theta$ with θ in (17) and the use of Bessel functions give comparable results.

Finally, lest we leave the reader in too much suspense, the narrator in Poe's story survives. He entices the dungeon rats to gnaw through his bonds, and escapes just as the pendulum had

> already pressed upon my bosom. It had divided the serge of the robe. It had cut through the linen beneath. Twice again it swung and a sharp sense of pain shot through every nerve. But the moment of escape had arrived I slid from the embrace of the bandage and beyond the reach of the scimitar. For the moment, at least I was free [**64**, p. 255].

Exercises

1. Suppose that a pendulum has length 4 feet. On a mountain top its period is 0.01 seconds longer than at sea level. Estimate the height of the mountain.

2. For the same pendulum as in Exercise 1, suppose that in a deep mine shaft the period of this pendulum is 0.01 seconds longer than at sea level. Estimate the depth of the mine.

3. Position a man at the top rung of a lightweight aluminum ladder of length 40 feet, so that the ladder can be modeled as the motion of an ordinary pendulum, with its mass at the end of the ladder. Take $k = 10$ ft/sec. Re-solve the ladder problem. Compare your solution with that of the problem of the text.

4. In the sliding ladder problem, (14) is somewhat surprising. Intuition might make one think that the moment of separation occurs when the tip of the ladder is accelerating downwards at $-g$ rather than $-3g/2$. At what magnitude is the center of mass of the ladder accelerating downwards at this moment of separation?

5. Design a ladder for which $y_c'' = -2g$ in (14).

6. Numerically solve (17) for the pendulum's velocity and generate a graph giving the pendulum's velocity with respect to time.

7. Suppose the forty-foot ladder of the chapter also shrinks (or collapses) at a constant rate as well as having its base moved outwards at a constant rate. Adjust the ladder model accordingly and determine when the ladder starts falling (rather than just sliding down the wall).

8. Model a forcing function, some kind of pump or impulse, so that Poe's pendulum does go from an amplitude of 0.5 feet at length 2 feet to an amplitude of 15 feet at 32 feet.

9. A swing is a modified pendulum. Describe a good algorithm for pumping a swing.

10. Model the motion of the Foucault pendulum and explain how it can be used to measure the earth's rotation.

preamble vi

Know then thyself,
Created half to rise, and half to fall.

Alexander Pope (1688–1744),
An Essay on Man

One year, my wife, Connie, our two teen-age sons, and I went to Tanzania, for me to teach as a Fulbright professor at the University of Dar es Salaam, and for all of us to have an adventure. I had chosen Tanzania because it was bounded by the warm waters of the Indian Ocean. Connie and I had been scuba divers in our college days. We'd promised each other that we would continue diving after marriage. However, living in Tennessee and raising children on a single income made that difficult.

Connie had been an avid diver. During her spring breaks before we married, she and various friends drove to Florida from New York, lived on oranges, and dove. My diving resumé was shorter. I had made only one dive, in a reservoir in Oklahoma on a cold spring morning. Water visibility was such that I could barely see my hands when held directly in front of my mask. I was less than an avid diver.

So twenty-some years after my only dive, the four of us signed up for courses on a beach in East Africa. While our sons took written exams for their beginner course, Hans, the dive master, took Connie and me out in a skiff for a checkout dive. The equipment that we donned was simpler than what I had used long ago. It all seemed too easy as we slipped into the waters. The skills of clearing a mask, equalizing ear-pressure, and breathing slowly and continuously came back. Our weight belts let us sink to the white sandy bottom, about thirty feet down. Here Hans had us practice taking off our equipment and putting it on again. After each drill, he signaled okay. I could see eighty feet in all directions. This was much better than a cold, muddy Oklahoma lake.

With growing confidence I experimented with my equipment controls. The buoyancy compensator device is a vest attached to the air regulator and air tank. One button on a hose-grip adds air to the vest, and another releases air to it. Thinking to add air in order to get a little more buoyant, I pushed a button. Air flowed in. I started rising more than I wanted. I pushed the other button, but nothing happened. So I pressed the first button again. It was the wrong choice, because more air entered my vest and I rose faster. I shot to the surface.

Fortunately, I remembered to breathe evenly, because otherwise my lungs might have ruptured. At the surface, I figured out what I'd been doing wrong and released the air in my vest. Connie and Hans, looking up, shook their heads in mock dismay, and beckoned with their hands for me to re-enter their world.

CHAPTER **VI**

Retrieving H. G. Wells from the Ocean Floor

In 1896, H. G. Wells wrote "In the Abyss" [**93**], a story about a differential equation. In it, a spherical capsule, that we call Maria (Wells refers to it only as the sphere), is attached to a heavy sinker with a cord 100 fathoms or 600 feet. The sinker falls towing Maria behind it, as in Figure 1. Once the sinker strikes the ocean floor, Maria slows her descent towards the bottom. When the time comes for Maria to ascend, she leaves her sinker behind and rises to the surface faster than her descent, breaking the surface and continuing upward to a great height before splashing down.

To give a flavor of the story, here are a few excerpts.

To start Maria's round trip,

> presently a man [Elstead] was to crawl in through that glass manhole, to be screwed up tightly, and to be flung overboard, and to sink down— down—down, for five miles.
>
> In fifty seconds everything was as black as night outside, except where the beam from his light struck through the waters, and picked out every now and then some fish or scrap of sinking matter.

When the sinker hits the ocean floor, Maria's descent slows abruptly.

> [Elstead] looked at the padded watch by the studs, and saw he had been traveling now for two minutes. Then suddenly the floor of the sphere seemed to press against his feet, the rush of bubbles outside grew slower and slower, and the hissing diminished. The sphere rolled a little.

When the cord is severed,

> abruptly the sphere rolled over, and he swept up, as an ethereal creature clothed in a vacuum would sweep through our own atmosphere back to

its native ether again. The sphere rushed up with even greater velocity than, when weighted with lead sinkers, it had rushed down.

When Maria erupts from the sea, as viewed from the surface support ship,

> a little white streak swept noiselessly up the sky, traveled more slowly, stopped, became a motionless dot, as though a new star had fallen up into the sky. Then it went sliding back again and lost itself amidst the reflections of the stars and the white haze of the sea's phosphorescence.

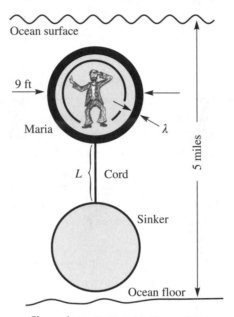

Figure 1. H. G. Wells's submersible

Wells' times for ascent and descent are not right. Of course, Wells really wasn't interested in specific numbers. He just wanted some numbers for his story. "It is only fiction," he may have rationalized. He has a character estimate that Maria should strike the ocean floor five miles down in 45 seconds, but this is a time almost as short as if the sphere was falling the same distance through a vacuum. Another character judges that the sphere experiences a constant acceleration of 2 ft/sec^2 during its five mile descent, but with that acceleration Maria would fall only about 2.8 miles in two minutes. The aquanaut Elstead estimates that by the twelfth second after launch, he will have descended about 75 feet. Such descent descriptions are incompatible. At other times, Wells is vague, as if sensing that he was getting his facts wrong. Rather than giving a specific time for the ascent, he merely says that it was faster than the descent. Finally, to break an inner tension which he must have been feeling as he continued groping for good-sounding numbers, he speaks through the tale's captain, espousing the philosophy, "I'm no slavish believer in calculations."

What Wells needed was a mathematical advisor. Just as Hesiod, he needed a muse. So let's model Maria's dynamics and see how we might improve upon Wells' numbers, while preserving the spirit of his story.

A falling model

In the story, Maria is a spherical shell of diameter nine feet made of "jolly thick steel." Let λ be its thickness. We assume that the mass of the tether of length L connecting Maria to its sinker is negligible. In Wells' story, L is 600 feet. We will find the minimum value of L so that Maria does not crash into the ocean floor. In the story, Wells describes the ballast at the end of the tether as "big lead sinkers," but for simplicity we shall assume that the ballast is a single solid lead sphere. Wells gives the weight of Maria and its sinker as a "dozen tons." We use the phrase *tandem spheres* to refer to Maria and her sinker. Let $R = 4.5$ ft be the radius of Maria, and let r be the radius of the sinker. We take the mass of one cubic foot of seawater as $\delta = 2$ slugs and weight $\delta g = 64$ pounds, of steel as $\epsilon = 15.1$ slugs, and of lead as $\gamma = 22$ slugs, where $g = 32$ ft/sec^2.

Because seawater cannot be compressed, a submerged object of constant volume experiences a constant buoyancy as it rises or falls. That is, if an object has mass m and volume V, its apparent mass m_a in seawater differs from m by the quantity of water it displaces, so that $m_a = m - \delta V$. The buoyancy force F_B on any submerged object is therefore $F_B = -m_a g$. If F_B is positive, negative, or zero, the corresponding mass is said to have positive, negative, or neutral buoyancy, respectively. Let M_μ and M_σ be the masses of Maria and its sinker, respectively, and let m_μ and m_σ be the apparent masses of Maria and its sinker. Since Maria is lighter than water, $m_\mu < 0$.

To complete our falling model, we must find specific values for λ and r. Since Maria floats,

$$\frac{4}{3}\pi \left(R^3 - (R - \lambda)^3\right) \epsilon = M_\mu < \frac{4}{3}\pi R^3 \delta \approx 763 \text{ slugs.} \qquad (1)$$

Solving (1) gives $\lambda < 2.5$ inches. If λ is close to 2.5 inches, then a submerged Maria will rise slowly. If λ is too small, Maria's hull will implode. To help select a good λ value in this range, we look at two early submersibles.

The first deep submersible was built in 1929 by Otis Barton, an engineer who was inspired by Wells' story, and William Beebe, a noted naturalist. In 1934, they logged a dive of 3000 feet in a hollow steel ball called a bathysphere. Its outside diameter was five feet and weighed 4500 pounds, making the hull about 1.5 inches thick. In the photo of the submersible in Figure 2, Beebe is on the left and Barton on the right. To descend, the ball was lowered by cable from a surface ship. Says Beebe about two men being inside the ball,

> I had no idea that there was so much room inside a sphere four and a half feet in diameter, although the longer we were in it the smaller it seemed to get [**9**, p. 657].

In 1960, the submersible Trieste descended to a depth of just under seven miles in the Marianas Trench. Trieste was built to withstand depths up to ten miles. Its spherical pressure hull was 6.6 feet in diameter and between 5 to 7 inches thick, and it was suspended under a large cylindrical tank filled with gasoline, a fluid lighter than water. Like Maria it was independent of a surface ship after launch. To sink, Trieste jettisoned gasoline, replacing it with sea water. To rise, it jettisoned ballast rocks. Unlike Maria, it took its time going down and up. Descending several miles took hours.

Figure 2. The Bathysphere, 1931, [**9**, p. 665], courtesy of the National Geographic Society

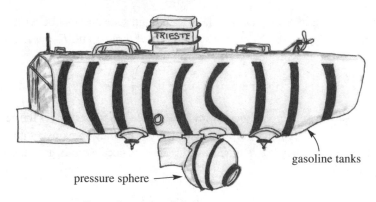

Figure 3. Sketch of Trieste, after [**17**, p. 232]

We will choose λ to be 1.5 inches. Even though this is too thin for Maria to survive the pressure of water five miles deep, we'll allow Wells this fiction. Maria then has a mass of about 467 slugs, which we round up to 500 slugs to account for the mass of an aquanaut and some gear. Her apparent mass is then $m_\mu = -263$ slugs, and her weight is eight tons.

The weight of seawater displaced by Maria alone is about 12.2 tons. In order for the tandem spheres to sink, they must weigh more than the water they displace. Therefore, we must increase Wells' estimate of twelve tons for the total weight of the tandem spheres. That is,

$$M_\mu + \frac{4}{3}\pi r^3 \gamma > \frac{4}{3}\pi (R^3 + r^3)\delta, \qquad (2)$$

which implies that the sinker's radius r must be at least 1.5 feet. Before we assign a value for r, let us explore Wells' description that Maria rises faster than she descends. To do so, we will need to derive her equations of motion.

The force F_D tending to slow an object passing through a medium is called drag. In our model we ignore turbulence, viscosity changes due to changing temperature or salinity, and ocean currents. In this case, it is known that the drag on an object varies directly with its horizontal cross-sectional area and with the square of its speed through the medium. (See Long [**50**] for a good discussion of this model.) So, $F_D = kAv^2$, where A is the cross-sectional area of the falling or rising object, v is the speed of the object, and k is a constant. When the tandem spheres are falling, then $A = \pi(R^2 + r^2)$, whereas when Maria falls or rises independently, $A = \pi R^2$, where $R = 4.5$ feet.

The total force F on an object falling through seawater can be written in two ways: $F = mv'$ by Newton's second law of motion and $F = F_B + F_D$. So, when the tandem spheres are moving downwards,

$$(M_\sigma + M_\mu)v' = -(m_\sigma + m_\mu)g + \pi k(R^2 + r^2)v^2. \tag{3}$$

F_D is positive since drag is an upward force. Once the sinker strikes the ocean floor, Maria moves downward independently. The drag remains an upward force. The buoyancy F_B is now positive as is F because v is growing less negative, so

$$M_\mu v' = -m_\mu g + \pi k R^2 v^2. \tag{4}$$

When Maria rises,

$$M_\mu v' = -m_\mu g - \pi k R^2 v^2. \tag{5}$$

In this case F_D is a downward force.

Now let $\kappa = \pi k R^2$. We can write (3), (4) and (5) as

Maria and sinker going down:

$$v' = \frac{(1 + (\frac{r}{R})^2)\kappa}{M_\sigma + M_\mu}\left(v^2 - \frac{(m_\sigma + m_\mu)g}{(1 + (\frac{r}{R})^2)\kappa}\right), \tag{6}$$

Maria going down: $v' = \dfrac{\kappa}{M_\mu}\left(v^2 - \dfrac{m_\mu g}{\kappa}\right),$ \hfill (7)

Maria going up: $v' = -\dfrac{\kappa}{M_\mu}\left(v^2 + \dfrac{m_\mu g}{\kappa}\right).$ \hfill (8)

Let v_Σ be the terminal downward speed of the tandem spheres, and v_μ be the terminal upward speed of Maria. Since these are attained when $v' = 0$, from (6) and (8),

$$v_\Sigma = -\sqrt{\frac{(m_\sigma + m_\mu)g}{(1 + (\frac{r}{R})^2)\kappa}} \quad \text{and} \quad v_\mu = \sqrt{-\frac{m_\mu g}{\kappa}}. \tag{9}$$

Wells describes the upward journey as taking less time than the descent, and since the spheres move with terminal speed through most of their five mile descent and ascent,

$$-v_\Sigma < v_\mu. \tag{10}$$

Since $m_\sigma = (\gamma - \delta)(4\pi r^3/3)$ and $m_\mu = -263$ slugs, (9) and (10) imply that $r < 1.9$ feet. So, in order for Maria to ascend faster than her descent, r must be smaller than 1.9 feet. For the tandem spheres to descend, r must be larger than 1.5 feet by (2). If we split the difference, and let r be 1.7 feet, then the apparent mass of the tandem spheres is about 148 slugs, whereas the tandem spheres displace 804 slugs of seawater. Define the float factor of a mass to be the ratio of its mass to the mass of the seawater it displaces. Objects with float factor exceeding 1 sink, whereas objects with float factor less than 1 float. If $r = 1.7$ ft, the tandem spheres have a float factor of 1.2. Such a small factor suggests that the terminal velocity for the tandem spheres should be fairly slow, nowhere near the 2.5 miles per minute of Wells' story. Increasing r to $r = 1.9$ ft is not much of an improvement.

Thus, to preserve Wells' descent time of two minutes, we must increase r beyond the bound as given by (10), which implies that Maria's ascent will be slower than her descent! (As r increases beyond 1.5 feet, the tandem spheres' descent time decreases, while Maria's ascent time remains constant. The break-even point is at $r = 1.9$ feet.) We therefore increase r substantially. For simplicity and symmetry, let r equal R.

The specific values we have assigned for the parameters in Wells' story are then $M_\mu = 500$ slugs, $m_\mu = -263$ slugs, $R = 4.5$ feet $= r$, $\lambda = 1.5$ inches, $M_\sigma \approx 8397$ slugs, $m_\sigma \approx 7634$ slugs. By (9), the ratio of the terminal velocities is

$$\frac{v_\mu}{v_\Sigma} = -\sqrt{-\frac{2m_\mu}{m_\sigma + m_\mu}} \approx 0.27, \qquad (11)$$

so the upward trip takes about four times as long as the downward trip. Therefore, Maria's ascent time must be approximately eight minutes, given that her descent time was two minutes.

Now let's determine κ. We can approximate it without solving (6). Since v_Σ is approximately the quotient of falling distance and falling time, by (9)

$$\sqrt{\frac{(m_\sigma + m_\mu)g}{2\kappa}} \approx \frac{5 \text{ miles}}{2 \text{ min}},$$

so $\kappa \approx 2.44$ slugs/ft.

To solve (6), we need an initial condition. We assume that $v(0) = 0$, even though in Wells' story, the sinker is freed from its tackle at the ocean's surface and is allowed to drop, gathering speed, until the tether is taut, whereupon Maria is dragged downwards. At time $t = 0$, we assume that Maria is at

the ocean's surface, the sinker is 600 feet down, and the tandem spheres
have zero speed. (We assume that the tether of negligible mass has unlimited
strength.)

With the initial condition $v(0) = 0$, the solution to (6) is

$$v(t) = -\sqrt{\frac{(m_\sigma + m_\mu)g}{2\kappa}} \tanh\left(\frac{t\sqrt{2\kappa g(m_\sigma + m_\mu)}}{M_\sigma + M_\mu}\right), \qquad (12)$$

where t is in seconds and $v(t)$ is in feet per second.

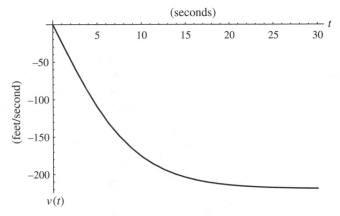

Figure 4. Reaching terminal velocity

From (12) the downward plunge of the tandem spheres attains terminal
velocity $v_\Sigma \approx -220$ ft/sec in about 15 seconds as in Figure 4. Assume that
at time $t = 0$, Maria's center is at the surface. Then the bottom of the sinker
is at $s(0) = -L - 3R$, where $s(t)$ is distance of the sinker's bottom from the
surface. From (12)

$$s(t) = \frac{M_\sigma + M_\mu}{2\kappa} \ln\left(\text{sech}\left(\frac{t\sqrt{2\kappa g(m_\sigma + m_\mu)}}{M_\sigma + M_\mu}\right)\right) - L - 3R. \qquad (13)$$

So after falling for two minutes, the sinker is 668 feet above the ocean
floor. Our estimate for κ was a little too high. Trial shows that a better
value for a descent time of two minutes is $\kappa \approx 2.31$. To find κ directly,
solve $s(2 \text{ min}) = 5$ mi using a computer algebra system. With this new κ
value, $s(2)$ is about seven feet short of five miles down, good enough for our
purposes. The updated terminal velocity is $v_\Sigma \approx -226$ ft/sec.

Finding the minimum tether length

To find the minimum length of the tether, we solve (7) with initial condition $v(0) = v_\Sigma$, getting

$$v(t) = \sqrt{-\frac{m_\mu g}{\kappa}} \tan\left(\frac{t\sqrt{-\kappa m_\mu g}}{M_\mu} + \tan^{-1}\left(v_\Sigma \sqrt{-\frac{\kappa}{m_\mu g}} \right) \right). \qquad (14)$$

(If the reader uses a computer algebra system, make sure that it knows that m_μ is negative, otherwise v may be given as a hyperbolic function.)

Integrate (14) so as to obtain distance $s(t)$ where $s(0) = 0$. Figure 5 gives a graph of $s(t)$. The graph is valid up to the turning point at $t \approx 4.7$ seconds, at which time Maria has traveled about 293 feet and come to a stop. Therefore, Wells can safely shorten his tether length from 600 to about 300 feet.

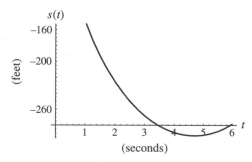

Figure 5. Maria gliding to rest

Into the air

The terminal velocity of Maria's ascent is about 60 ft/sec by (9). If we neglect air resistance, then Maria's velocity from the time she breaks the surface to splash-down is $v(t) = -32t + 60$ and her height above the surface is $s(t) = -16t^2 + 60t$. Maria's extremal height of 56 feet occurs in 1.875 seconds. After cresting, Maria crashes into the sea with speed 60 ft/sec. Elstead probably will not survive such an impact. Thus Maria should be redesigned so that she rises more gently to the surface, at least in the last few seconds of her ascent. Two ideas for doing so are given in Exercises 6 and 8.

Conclusions

Do the above errors—the incompatibility of Wells' descent and ascent times, the thinness of Maria's hull, the overly fast upward terminal velocity—ruin Wells' story?

For most readers, the answer is no. Most readers are interested only in the drama. For readers who know some calculus, Wells' errors make the story better, as they cause us to rethink our ideas of how objects fall. Of course, leaving a lead sinker with diameter nine feet behind each time Maria makes a dive is logistically inefficient. Besides that, the undersea civilization that Elstead discovers in Wells' story doesn't appreciate such items raining down upon them.

Wells' story gives a new version of the problem of bailing out of airplanes and opening parachutes, that is, of determining terminal velocities and descent times. After revising some of his numbers, his story gives a good example of how objects fall when encountering resistance near the earth's surface.

Exercises

1. Find the time to descend five miles if Elstead's estimate, that after 12 seconds Maria is at a depth of 75 feet, is accurate.

2. When Wells wrote his story no one had ever gone deep enough (and returned to talk about it) to verify when sunlight fails. Barton and Beebe in their bathysphere found that on very bright days, sunlight can penetrate to almost 2000 feet [53, p. 85]. Elstead estimated that "in fifty seconds everything was as black as night outside." Given that at fifty seconds, Maria is at 2000 feet, estimate how long it takes Elstead to reach the ocean floor.

3. Resolve the problem of this chapter given that Maria's hull thickness is $\lambda = 1.5$ inches, constructed from a steel alloy with density $\rho = 20$, while the sinker is made of ordinary steel, $\epsilon = 15.1$.

4. Suppose a solid steel ball of diameter two feet has terminal velocity through seawater of 100 feet per second. Use this information to determine how long it should take Maria and her sinker to descend five miles.

5. Find the tension in the tether between Maria and her sinker on descent.

6. Imagine that Maria can deploy a fin-like skirt to increase her surface area as depicted in Figure 6. Determine the dimensions of this fin so as to reduce Maria's terminal velocity to 10 ft/sec. When should this fin be deployed? (Assume Maria also deploys an anchor to maintain stability.)

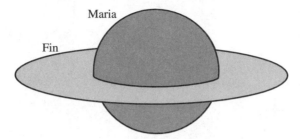

Figure 6. Maria and fin

7. Generalize (11) in terms of r_σ, r_μ, and λ, where r_σ is the radius of the sinker, r_μ is the outside radius of Maria, and λ is the thickness of Maria's hull. Use this to design a capsule-sinker system in which the capsule's ascent time is faster than its descent time.

8. Here's a way for Maria to slow her ascent. Suppose Maria takes on water at the rate of 1 cubic foot each second as she ascends. Will she make it to the surface from five miles deep? If not, at what point should she begin taking on water at this rate? Determine the rate ρ so that if Maria begins taking water on at rate ρ at depth five miles she reaches the surface with speed 10 ft/sec.

9. If Maria's hull had been five inches thick, determine the diameter needed for her to rise from the ocean floor. Then find the diameter so that the five mile ascent trip lasts thirty minutes.

10. Suppose that the drag force varies directly with diameter and velocity so that (6) becomes

$$v' = -\frac{k(D + d)}{M_\sigma + M_\mu} \left(v + \frac{(m_\sigma + m_\mu)g}{k(D + d)} \right),$$

where D is the diameter of Maria, d is the diameter of the sinker, and k is some positive constant. Determine Maria's speed and position and compare the conclusions reached with those from the other model.

preamble vii

By heaven, man, we are turned round and round in this world,
like yonder windlass, and Fate is the handspike.
Herman Melville (1819–1891), *Moby Dick*

Friends from England were coming to Tennessee for a visit. When we had visited them years ago, they had taken us on a hike on empty moors, along an old canal route, and into a village pub. We needed a similar outdoor event with plenty of activity to counteract the confinement of the airplane and rental car, especially for their three children. But it couldn't be too strenuous because we wanted to enjoy our time together. Why not cycle down Whitetop Mountain?

Our home is in Appalachia, where rolling green mountains stretch from southern Georgia on up to Maine. When the first railroads were built in this territory, tracks followed mountain streams and cut across high passes. The steam engines puffed and chugged, crept uphill, and rode their brakes downhill. At Whitetop, the rails are gone, but the trail remains, called the *Creeper* in honor of the old engines.

The part of the trail we chose was a seventeen mile stretch from the crest of the mountain down to the village of Damascus. Here we could rent bicycles and be driven up to the mountain top. A splendid plan! Nothing could go wrong. All we had to do was roll downhill. To save money, we used a few of our own bicycles, which turned out to be a mistake.

The nine of us talked and laughed our way up the mountain, reminiscing and catching each other up on what had happened in the years since we'd last been together. At the top the bicycle shopkeeper wished us a pleasant journey. In two or three hours we should see him again in Damascus. We mounted our cycles and coasted downhill. The sun was shining, a gentle cool breeze was blowing, birds sang in the trees. Life was great.

Then my front tire bumped over a narrow crevice across the trail. The wheel folded in half. I tumbled over the handlebars, landing in a heap with the bike atop me. I got up, dusted myself off. The nearest replacement wheel

97

was seventeen miles away. Cell phones had not yet been invented. Few people frequented this part of the trail on week days. What to do? In frustration, I picked up the bike, raised it over my head, and heaved it into the bushes. I wasn't coming back for it.

I sent everyone on ahead to glide down the mountain. Meanwhile I walked and jogged. It only took six hours.

CHAPTER VII

Sliding along a Chord through a Rotating Earth

Connect any two cities on earth's surface by a straight-line subway track. If a train, powered only by gravity, runs along without resistance, how long will it take the train to complete its route, taking into account the earth's rotation? A bit of linear algebra handles this twist to a classic problem and gives some surprising answers.

We solved a special case in Chapter II, when we dropped a stone down a shaft at one of the poles of the earth. When the shaft is an arbitrary chord of a nonrotating earth, the problem appears in many introductory physics texts such as [**71**, pp. 258–9] and recreational mathematics books such as [**8**]. The idea is natural. Arthur Conan Doyle, the author of the Sherlock Holmes stories, used the idea in one of his early stories in which two English diamond miners in South Africa describe a deep fissure in the earth:

> It's my belief that it [the hole] extends through to old England. 'Twould save postage if we could drop our letters through [**26**, p. 263].

Readers unfamiliar with vector and matrix manipulation may find items **2**, **4**, **5**, and **10** of the appendix helpful in following the next few paragraphs.

Let us position earth's center at the origin O so that the equator lies in the xy-plane and the north pole is on the positive z-axis. Let M be the rotation matrix

$$M = M(t) = \begin{bmatrix} \cos\phi & -\sin\phi & 0 \\ \sin\phi & \cos\phi & 0 \\ 0 & 0 & 1 \end{bmatrix},$$

where $\phi = \phi(t) = 2\pi t/Q$, Q is the period of the earth about its axis, and t is time. Let $\hat{A}$ and $\hat{B}$ be any two cities on earth's surface. Let A and B be the positions of the two cities at $t = 0$. Let $\mathbf{u}$ be the unit vector in the $A - B$ direction, C the segment joining $\hat{A}$ and $\hat{B}$, $\hat{N}$ the point on C nearest to O, N the position of $\hat{N}$ at $t = 0$, and $\mathbf{w}$ the unit vector pointing from the origin to N. (If $\hat{A}$ and $\hat{B}$ are at the antipodes, let $\mathbf{w}$ be any unit vector perpendicular to $\mathbf{u}$.) When earth's surface is a sphere, $\hat{N}$ is the midpoint of C. The vectors $\mathbf{u}$ and $\mathbf{w}$ are orthogonal. Condense the train to a single point. Let P be the position of the train along C at time t, and denote by c, r, and s the respective distances from O to $\hat{N}$, from O to P, and from $\hat{N}$ to P. We suppose s is positive when P is between $\hat{N}$ and $\hat{A}$ and negative when P is between $\hat{N}$ and $\hat{B}$. We take vectors and positions as being in column form. As the earth rotates under the isometry M, the two cities are at the positions MA and MB; the vectors $M\mathbf{u}$ and $M\mathbf{w}$ remain orthogonal vectors and point in the directions from $\hat{B}$ to $\hat{A}$ and from O to $\hat{N}$, respectively (see Figure 1). Unless otherwise stated, we assume that earth is a sphere of radius $R = 6400$ km and that the acceleration due to gravity at position $\mathbf{r}$ is $\mathbf{f}(\mathbf{r}) = -k\mathbf{r}$ where $k = g/R$, $g = 9.8$ m/sec^2, and $|\mathbf{r}| \leq R$.

The train's position is given by

$$P = cM\mathbf{w} + sM\mathbf{u}. \tag{1}$$

Since c, $\mathbf{u}$, and $\mathbf{w}$ are constant, the velocity $\mathbf{v}$ and the acceleration $\mathbf{a}$ of the train are

$$\mathbf{v} = cM'\mathbf{w} + s'M\mathbf{u} + sM'\mathbf{u}$$

and

$$\mathbf{a} = cM''\mathbf{w} + s''M\mathbf{u} + 2s'M'\mathbf{u} + sM''\mathbf{u}, \tag{2}$$

where

$$M' = \frac{2\pi}{Q} \begin{bmatrix} -\sin\phi & -\cos\phi & 0 \\ \cos\phi & -\sin\phi & 0 \\ 0 & 0 & 0 \end{bmatrix},$$

$$M'' = \left(\frac{2\pi}{Q}\right)^2 \begin{bmatrix} -\cos\phi & \sin\phi & 0 \\ -\sin\phi & -\cos\phi & 0 \\ 0 & 0 & 0 \end{bmatrix}.$$

The acceleration acting on the train along the chord, given by (2), is the projection of $\mathbf{a}$ onto the unit vector $M\mathbf{u}$, or $\mathbf{a} \cdot M\mathbf{u}$. For the moment, let us think that the only acceleration in this system is that due to gravity. Then, the acceleration along the chord should also be given by the projection of $-k\mathbf{r}$

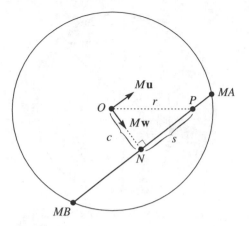

Figure 1. A singular train sliding along a chord in a rotating earth

onto the chord. That is, the acceleration of the train along the chord should
satisfy

$$(cM''\mathbf{w} + s''M\mathbf{u} + 2s'M'\mathbf{u} + sM''\mathbf{u}) \cdot M\mathbf{u} = -ks. \tag{3}$$

Let T denote the transpose operator. Since the dot product $\mathbf{x} \cdot \mathbf{y}$ is $\mathbf{x}^T\mathbf{y}$ and
since $(XY)^T = Y^T X^T$, (3) becomes

$$s'' - \left(\frac{2\pi}{\varrho}\right)^2 (c\mathbf{w}^T + s\mathbf{u}^T) \begin{bmatrix} 1 & 0 & 0 \\ 0 & 1 & 0 \\ 0 & 0 & 0 \end{bmatrix} \mathbf{u} = -ks, \tag{4}$$

because $M^T M = I$, the identity matrix, $(M'\mathbf{u})^T M\mathbf{u} = 0$, and

$$(M'')^T M = -\left(\frac{2\pi}{\varrho}\right)^2 \begin{bmatrix} 1 & 0 & 0 \\ 0 & 1 & 0 \\ 0 & 0 & 0 \end{bmatrix}.$$

We know that the acceleration on the train should be the sum of its gravitational acceleration and its centripetal acceleration. The centripetal acceleration on a particle at position P is $\omega^2\mathbf{d}$, where ω is the rotation rate and
$\mathbf{d}$ is the vector of displacement of the particle from the axis of rotation (see
Figure 2).

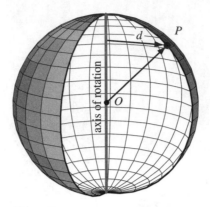

Figure 2. Cutaway of a spherical shell

For the train, $\omega = 2\pi/Q$ and

$$\mathbf{d} = \begin{bmatrix} 1 & 0 & 0 \\ 0 & 1 & 0 \\ 0 & 0 & 0 \end{bmatrix} P.$$

Thus the projection of $\omega^2 \mathbf{d}$ onto the direction of the tunnel is

$$\omega^2 \mathbf{d} \cdot M\mathbf{u} = \left(\frac{2\pi}{Q}\right)^2 \begin{bmatrix} 1 & 0 & 0 \\ 0 & 1 & 0 \\ 0 & 0 & 0 \end{bmatrix} (cM\mathbf{w} + sM\mathbf{u}) \cdot M\mathbf{u},$$

which simplifies to

$$\left(\frac{2\pi}{Q}\right)^2 (c\mathbf{w}^\mathsf{T} + s\mathbf{u}^\mathsf{T}) \begin{bmatrix} 1 & 0 & 0 \\ 0 & 1 & 0 \\ 0 & 0 & 0 \end{bmatrix} \mathbf{u},$$

a term in (4). That is, (4) says, as it should, that $\mathbf{s}''$ is the sum of gravitational and centripetal acceleration.

The form of (4) is

$$s'' + \lambda s = \mu, \tag{5}$$

where λ and μ are constants. If $\mathbf{u} = (u_1, u_2, u_3)$ and $\mathbf{w} = (w_1, w_2, w_3)$, then

$$\lambda = k - \frac{4\pi^2}{Q^2}\left(u_1^2 + u_2^2\right), \quad \mu = \frac{4c\pi^2}{Q^2}(u_1 w_1 + u_2 w_2). \qquad (6)$$

When λ is positive, the solution of (5) is

$$s = \frac{\mu}{\lambda} + \alpha \cos(\sqrt{\lambda} t + \beta), \qquad (7)$$

where α and β are constants. (See the entry on simple harmonic motion in the appendix for details in solving the equation.) Since the train leaves earth's surface from rest we have the initial conditions $s(0) = \gamma$ and $s'(0) = 0$, where γ corresponds to the signed distance from N to city $\hat{A}$. It is not hard to see that $\beta = 0$ and $\alpha = \gamma - \mu/\lambda$, so (7) becomes

$$s = \frac{\mu}{\lambda} + \left(\gamma - \frac{\mu}{\lambda}\right)\cos(\sqrt{\lambda}\, t). \qquad (8)$$

If the train ventures beyond the surface of the earth under the motion described by (8), the equation no longer models the train's behavior. If $\mu \neq 0$, then the simple harmonic motion of (8) results either in the train failing to reach city $\hat{B}$ before returning to city $\hat{A}$ or in the train proceeding beyond city $\hat{B}$, when it becomes a projectile. At $t = 0$ the train is in city $\hat{A}$, which is at distance γ from N. At $t_1 = \pi/\sqrt{\lambda}$, the train will be at its opposite extreme, $-\gamma + 2\mu/\lambda$. Thus the train's shortfall or surplus distance in reaching city $\hat{B}$ is given by $2\mu/\lambda$. We explore this in Example 3.

With respect to Q and the two cities, by (6), the period $2\pi/\sqrt{\lambda}$ of (8) ranges from a minimum of $2\pi/\sqrt{k}$ to a maximum of $2\pi/\sqrt{k - 4\pi^2/Q^2}$. The train's period increases as the rotation rate of earth quickens, which makes sense because, as the rotation rate of a body increases, the centripetal acceleration eventually overcomes the gravitational acceleration at any point not on the axis of rotation. If the gravitational and centripetal accelerations in the direction of the tunnel sum to zero for a train at its start, the train fails to slide down the tunnel. If gravitational acceleration slightly exceeds centripetal acceleration, the train proceeds down the tunnel with a long period.

When $\lambda = 0$, the solution to (5) is $s = \mu t^2/2 + \gamma$, which occurs only if earth's rotation rate is speeded up dramatically. When λ is negative, the solution is

$$s = \frac{\mu}{\lambda} + \left(\gamma - \frac{\mu}{\lambda}\right)\cosh(t\sqrt{-\lambda}). \qquad (9)$$

If $\gamma \neq \mu/\lambda$, then the train ultimately flies off into space. To demonstrate that the train may not immediately fly off the earth (for $\lambda < 0$), see Example 4.

We now consider some specific chords.

Example 1. Chords parallel to the z-axis. Here $\mathbf{u} = (0, 0, 1)$, so that (4) becomes

$$s'' = -ks.$$

Trains sliding along these chords follow simple harmonic motion with period $2\pi/\sqrt{k}$, which is about 84.6 minutes. This is the same as that for any chord for a nonrotating earth.

Example 2. Chords lying in the equatorial plane. For any chord in the equatorial plane, the vectors $\mathbf{u}$ and $\mathbf{w}$ both have third components 0, so (4) becomes

$$s'' = \left(-k + \frac{4\pi^2}{Q^2}\right)s.$$

This gives simple harmonic motion with a period $2\pi/\sqrt{k - 4\pi^2/Q^2}$, if the radical is real. For our earth, where $Q = 24 \cdot 60 \cdot 60 = 86{,}400$ seconds, the period is about nine seconds longer than along the chords in Example 1.

If the period of earth is shortened, the period for the train increases along chords in the equatorial plane. For example, if Q is 2 hours, then the period of the train is about 119 minutes. Shorter periods Q may result in the train, as well as everything else, flying from earth's surface.

Example 3. Chords between the south pole and the equator. We take $\hat{A}$ to be Entebbe, Uganda, and $\hat{B}$ as the south pole. Without loss of generality, let $A = (R, 0, 0)$. Let $B = (0, 0, -R)$. Thus $\mathbf{u} = (1, 0, 1)/\sqrt{2}$, $\mathbf{w} = (1, 0, -1)/\sqrt{2}$, and $c = R/\sqrt{2}$. If the train starts at A, the initial conditions are $s(0) = R/\sqrt{2}$ and $s'(0) = 0$. With these values, $2\mu/\lambda > 0$, so the polar express turns back $2\sqrt{2}\pi^2 R/(kQ^2 - 2\pi^2) \approx 15.7$ km short of the south pole.

On the other hand, if the train starts at the south pole, it glides on past Entebbe, becomes a projectile, and crashes about 22 km to the north.

Example 4. On a pulsar. Take $Q = 1$ second, the rotation rate of a typical pulsar. With R and g as before, let $A = (0, 0, -R)$ and $B = (R, 0, 0)$, so $\mathbf{u} = (-1, 0, -1)/\sqrt{2}$, $\mathbf{w} = (1, 0, -1)/\sqrt{2}$, $c = R/\sqrt{2} = \gamma$. Then $\lambda = k - 2\pi^2 \approx -2\pi^2$, $\mu = -\sqrt{2}\pi^2 R$, and $\gamma - \mu/\lambda \approx -0.000001875$ meters. By (9),

$$s \approx \frac{R}{\sqrt{2}} - 0.000001875\cosh(\pi t\sqrt{2}).$$

Figure 3. The Polar Express

That is, the train proceeds along the chord with ever increasing speed until it erupts from the pulsar at the equator at $t \approx 6.73$ seconds, traveling at about 40,000 km/sec. Figure 3 is a whimsical sketch of such a train ride, not drawn to scale. If the train had started at B instead, it would have immediately flown away from the earth. (If $Q = 1$, the earth is unstable, but we can imagine that its material is super-glued together.)

Example 5. The polar express for earth. The earth is not a perfect sphere, but is an oblate spheroid, bulging at the equator by $\delta \approx 7.1$ km and flattened at the poles by $\epsilon \approx 14.2$ km. If we assume that earth's gravitational acceleration is approximately $\mathbf{f}(\mathbf{r}) = -g\mathbf{r}/R$ as before then we can use the model. Take $R \approx 6371$ km, so that at $t = 0$, Entebbe is at $A = (R + \delta, 0, 0)$ and the south pole is at $B = (0, 0, -R + \epsilon)$. Then N is no longer the midpoint of $\overline{AB}$. In the xz-plane, the line through A and B is given by

$$\frac{z}{x - (R + \delta)} = \frac{R - \epsilon}{R + \delta},$$

while the line from the origin through N has the equation

$$z = \frac{(R + \delta)x}{\epsilon - R}.$$

Solving these two equations for the x-coordinate of N gives

$$x = \frac{(R + \delta)(R - \epsilon)^2}{(R + \delta)^2 + (R - \epsilon)^2}.$$

Thus N's coordinates are $(3178.4, 0, -3189.0)$. The distance from N to A is $\gamma \approx 4517.6$ km, the distance from N to B is $\Gamma \approx 4487.4$ km, and $2\mu/\lambda \approx 15.5$ km. When the train starts at Entebbe, it wants to turn back about 15.5 km short of $-\gamma$. But $\gamma - \Gamma \approx 30.2$ km, which means that the train wants to turn back about 14.7 km beyond the south pole!

   In the next chapter we solve a related problem. Instead of dropping a pebble (or a train) down a straight-line hole, we will allow it to fall as it will and make its own hole. We will see that the results will be simple harmonic motion again, but not in a straight line!

Exercises

1. Solve Doyle's problem as posed in the second paragraph of this chapter. That is, find the delivery time for a letter sliding from Kimberly, 24°E 28°S, to London, 0°E 52°N. Use the spherical earth model. How close to London will the letter approach? For a letter going from London to Kimberly, find its velocity at Kimberly.

2. For the polar express of Example 3, find an initial velocity v_o for the train at Entebbe so that it will arrive at the south pole with zero velocity.

3. Prove that chords corresponding to the maximal period have endpoints at the same latitude.

4. In Example 5, we took $\mathbf{f}(\mathbf{r}) = -g\mathbf{r}/R$. Design a function $\mathbf{f}(r, \phi)$ giving the gravitational acceleration at radial distance r from the origin and latitude ϕ, so that for any ϕ, $\mathbf{f}(r, \phi)$ is linear in r and so that for any point (r, ϕ) along the prime meridian at the earth's surface, $\mathbf{f}(r, \phi)$ is

approximately the earth's surface gravity at latitude ϕ and longitude 0. Find the polar express route using $\mathbf{f}(r, \phi)$. Contrast your results with those in Example 5.

5. Instead of a chord with endpoints A and B, take a circular arc in the equatorial plane with endpoints A and B (whose center will lie above the earth). Find the period for a pebble sliding along the arc. For which of these arcs is the period least?

6. Rather than assuming that gravitational acceleration is $\mathbf{f}(\mathbf{r}) = -g\mathbf{r}/R$, use a function representative of our earth as given in Figure 5e of Chapter II, and re-solve the problem of this chapter. Characterize the differences in the solutions between these two models.

7. Design a continuous function $\mathbf{f}(r, \phi)$ giving the gravitational acceleration at radial distance r from the earth's center and at latitude ϕ so that for the chord connecting Entebbe (on the equator) to the south pole, a train starting from Entebbe at zero velocity reaches the south pole with zero velocity.

8. Assume that gravity is constant and that the earth does not rotate. Given two points, find the curve through them down which an object's resistance-free sliding time is least.

9. Solve the brachistochrone problem of Exercise 8 if $\mathbf{f}(\mathbf{r}) = -g\mathbf{r}/R$.

10. Now solve the brachistochrone problem of Exercise 9 if the earth rotates.

preamble viii

After solving the equations of motion in this next chapter for a pebble falling freely through the earth, I saw that a pebble falls further south than east when dropped in northern climes.

Somewhere I must have made a mistake, I thought. After checking and rechecking, I set the work aside, vowing to return later with a fresh mind. Meanwhile, I ran across an account in which Newton describes an experiment of dropping a ball from about 30 feet to see how far east and south the ball lands from straight down. Newton estimated that the disparity was measurable.

I resolved to try it. My colleague Ray and my son Chris, home on break from graduate studies in mechanical engineering, volunteered to help. Within the winding stair-well of our four-story science building is a narrow, vertical drop space of fifteen meters. At the top, we secured a pulley. We threaded a fish line through a small steel ball with a fine hole drilled through its center. Using the pulley, we lowered the ball to a box of sand on the basement floor, and marked an X where the ball came to rest. Upon this X we centered a paper bullseye. Next, with the ball dangling just below the pulley, we used Newton's suggestion of severing the string with a lit match, thus limiting any extraneous initial velocity of the ball. Although the melting line induced some rotation to the ball, the ball soon dropped. It tore through the paper target below, scattering sand everywhere.

Upon inspection, my heart sank. The ball had passed through the center of the bullseye. The abberation was basically zero. Although one trial in an experiment is not definitive, I had seen enough.

As I thought about the logic of the theoretical solution and the care we'd taken in the experiment, I realized that the calculations were right and that the experiment was accurate. My understanding of what physicists meant by down was wrong. It's not the direction along a straight line from the drop site to earth's center! We explore these matters in the next chapter.

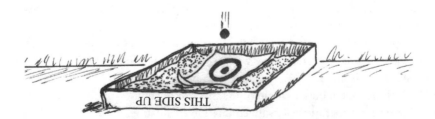

CHAPTER VIII

Falling through a Rotating Earth

Drop a pebble into a hole in the earth. Better yet, allow the pebble to drill its own hole as it falls. Along what path will it fall? How close to the center of the earth will it approach? How fast will it do so? Keep in mind that the earth rotates. Ignore all issues involving resistance.

Before answering these questions, we give some history behind this classic problem. Then we derive the differential equation governing the pebble's motion when dropped from the surface of the rotating earth. Assuming a linear gravitational field within the earth, we generate analytic solutions, showing that the pebble's path is an ellipse whose center is the center of the earth. When the pebble is dropped at the equator it misses the center by over 300 km. In doing so, the pebble moves in accordance with a familiar parametrization of the ellipse. Once we know the pebble's path, we generate the pebble's actual route through the earth, and then solve the equation using other gravitational fields for the earth.

A little history

Deep holes elicit mystery. For who among us when encountering a well with shadowed bottom is not tempted to drop a pebble and wait, listening for a splash? Witness the romance of the deep hole in popular literature, such as Alice falling down a rabbit hole for ever so long that she grows sleepy, and thinks she must surely have long since fallen beyond the earth's center [18, pp. 12–4], or the more recent Gandalf the Grey of *The Lord of the Rings* falling down an abyss that "none has measured," falling down unto "the uttermost foundations of stone," [87, p. 490]. Figure 1 shows Gandalf's trajectory through the Mines of Moria had he kept falling without banging into the sides of the hole.

Figure 1. Gandalf falling through Middle Earth, [76, cover], courtesy of Jason Challas and Lauren Gregory

Natural philosophers have considered this problem for millennia. For their thought experiments, they have used a variety of objects.

anvils: 700 BC. Hesiod claimed that an anvil would fall for nine days before reaching the underworld [40, lines 724–5, p. 131].

water: 399 BC. Socrates imagined that the earth was honeycombed with chasms, the largest of which "is bored right through the whole earth," and that water cascaded in oscillatory fashion falling from the earth's surface down to the earth's center and rising again, Plato [62, p. 383].

clods, chunks of earth: 350 BC. In explaining why the earth is spherical, Aristotle imagines that a clod "to travel towards the [earth's] center must prevail [fall] until it possesses the center with its own center" [5, book ii, chapter ii, p. 251].

boulders: first-century AD. Plutarch pointed out that if the earth is spherical, then boulders of 40 tons, like those thrown up by some Mediterranean volcanoes of his day, should they fall through the earth would pass through the earth's center and beyond and then return again, "oscillating in an incessant and perpetual see-saw" [63, p. 65].

a. Discussing the hole   b. Working in the hole

Figure 2. Thinking and doing

Satan: 1300. When Dante's Satan is cast from Heaven in *The Divine Comedy* he plunges into the earth, so hollowing out a void called Hell as he slows to zero velocity, winding up mired at the earth's literal center, forever doomed to munch on the shades of various historical villains (canto xxxiv).

millstones: 1624. Henrik van Etten, in one of the first books of recreational mathematics problems, argued that a millstone dropped down a hole would reach the earth's center in 2.5 days, whereupon "it would hang in the air" [**89**]. See Singmaster [**81**] for an exhaustive bibliography of early works, including van Etten, that mention the idea of falling down a hole through the earth.

cannonballs: 1632. Galileo analyzed the simple harmonic motion of a cannonball dropped down a perforation through the earth [**35**, p, 227]. Figure 2a is the frontispiece from Galileo's great dialogue, showing his three debaters arguing about how objects move in space.

balls: 1679. In an exchange of letters between Isaac Newton and Robert Hooke, Newton described an experiment using a falling ball to prove that the earth rotates. Newton speculated that should the ball fall through the earth it would spiral around to the earth's center. Hooke replied that in the absence of resistance the trajectory should be a closed elliptical-like loop with the earth's center as the center of the loop. See Arnol'd [**6**, pp. 15–26] for a more detailed account of this correspondence.

black holes: 1988. As a potential power source, Stephen Hawking drops a black hole, with mass the size of a mountain and whose radius is that of the nucleus of an atom, from the earth's surface [**37**, pp. 108–9].

While these drops were purely speculative, a French astronomer, in 1909, advocated construction of a deep hole in England (not France!), using convicts and peace-time armies for labor; Figure 2b [**32**, p. 351] is a fanciful view of this hole as work is progressing. About such an enterprise (as proposed earlier in 1750), Voltaire cautioned that the mouth of the hole would stretch "about three or four hundred leagues of earth, a circumstance that might disorder the balance of Europe," and suggested that the proposer of this idea "might need to conceal himself [therein] from the disgrace to which the publication of such absurd principles has exposed him" [**92**, pp. 195–6].

Deriving the differential equation

We wish to drop a pebble P at the equator of a rotating earth whose density is spherically symmetric. As we did in Chapter IV, we position the polar plane through the equator with earth's center at the origin so that the north pole is above the plane, as in Figure 3. With respect to pebble P's position at the point (r, θ), the two natural orthogonal unit vectors are $\mathbf{u}_r = \cos(\theta)\,\mathbf{i} + \sin(\theta)\,\mathbf{j}$

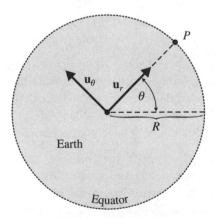

Figure 3. The natural directions

and $\mathbf{u}_\theta = -\sin(\theta)\,\mathbf{i} + \cos(\theta)\,\mathbf{j}$ and P's acceleration $\mathbf{a}$ is given by

$$\mathbf{a} = \left(r\frac{d^2\theta}{dt^2} + 2\frac{dr}{dt}\frac{d\theta}{dt}\right)\mathbf{u}_\theta + \left(\frac{d^2r}{dt^2} - r\left(\frac{d\theta}{dt}\right)^2\right)\mathbf{u}_r. \qquad (1)$$

Assume that pebble P's only acceleration is that of the gravitational field of the earth. Let $f(r)$ be the gravitational acceleration function induced by earth's mass at r units from its center. In (5), such an assumption constrains the $\mathbf{u}_\theta$ coefficient to be 0 and the $\mathbf{u}_r$ coefficient to be $f(r)$, so

$$0 = r\frac{d^2\theta}{dt^2} + 2\frac{dr}{dt}\frac{d\theta}{dt} = \frac{1}{r}\frac{d\left(r^2\frac{d\theta}{dt}\right)}{dt} \quad \text{and} \quad f(r) = \frac{d^2r}{dt^2} - r\left(\frac{d\theta}{dt}\right)^2. \qquad (2)$$

The first of the equations in (8) is the law of the conservation of angular momentum; that is, since $1/r$ is never 0,

$$r^2\frac{d\theta}{dt} = h, \qquad (3)$$

where h is a constant. In terms of h, the second equation in (8) can be written as

$$\boxed{f(r) = \frac{d^2r}{dt^2} - \frac{h^2}{r^3}.} \qquad (4)$$

General differential equation for a falling pebble

Differential equation (4) models the motion of a small particle moving freely in a fixed body's gravitational field. In the absence of friction and any other forces, the falling pebble's path is found by solving (4) along with the initial conditions,

$$\boxed{r = R \text{ when } t = 0, \ \theta = 0, \ \frac{dr}{dt}\Big|_{t=0} = 0, \ \frac{d\theta}{dt}\Big|_{t=0} = \frac{2\pi}{Q},} \qquad (5)$$

Initial conditions for (4)

where R is the radius of the earth and Q is the period of one revolution of the earth about its axis. Because the pebble is dropped at the equator, the initial rotation rate is that of the earth, $\frac{d\theta}{dt}\big|_{t=0} = 2\pi/Q$. By (5) the constant h of (3) is given by

$$\boxed{h = \frac{2\pi R^2}{Q}.} \qquad (6)$$

Constant of angular momentum

Since we wish to plot P's path in polar coordinates, a useful trick to solve (4) for r in terms of θ is to let $z = 1/r$. Then, as in Chapter IV,

$$\frac{d^2z}{d\theta^2} + z = -\frac{1}{h^2z^2} f\left(\frac{1}{z}\right). \tag{7}$$

Alternate general differential equation with respect to (4)

The Linear Model

Imagine a homogeneously dense earth, which is a simple and natural model. With it, Newton derived the corresponding gravitational force function culminating in corollary iii of proposition 91 of book i of the *Principia* as follows.

Think of the earth as composed of concentric spheres. Newton showed that the net gravitational attraction of any sphere on any body located anywhere inside the sphere is zero and that the net gravitational attraction of any sphere on any body located anywhere outside the sphere is exactly the same as that of a point of identical mass located at the center of the sphere. So as a body passes through the earth, the only part of the earth that attracts it consists of those spheres whose radii are less than or equal to the distance of the object from the center of the earth. Thus the mass acting on the body is proportional to the cube of its distance from the center of the earth. The force is proportional to the mass and inversely proportional to the square of the distance, and so the force $f(r)$ is directly proportional to the distance, giving $f(r) = -kr$, where k is some positive constant. (See Chapter II for a more formal argument.)

With $f(r) = -kr$, (4) and (7) become

$$\frac{d^2r}{dt^2} + kr = \frac{h^2}{r^3} \quad \text{and} \quad \frac{d^2z}{d\theta^2} + z = \frac{k}{h^2z^3}. \tag{8}$$

The twin differential equations for a pebble falling through the earth

Both of these nonlinear differential equations are of the form

$$\frac{d^2w}{du^2} + \alpha^2 w = \frac{\beta^2}{w^3}, \tag{9}$$

where α and β are constants. Since we wish to solve (8) for the initial conditions, $r(0) = R = 1/z(0)$, $\frac{dr}{dt}|_{t=0} = 0$ and $\frac{dz}{d\theta}|_{t=0} = 0$, the initial conditions for (9) are $w(0) = \delta$ for some $\delta > 0$ and $\frac{dw}{du}|_{t=0} = 0$. To solve this equation, we use the method of reduction in order, letting $\rho = \frac{dw}{du}$, which means that $\frac{d^2w}{du^2} = \frac{d\rho}{du} = \frac{d\rho}{dw}\frac{dw}{du} = \rho\frac{d\rho}{dw}$. Hence (9) becomes

$$\rho\, d\rho = \left(\frac{\beta^2}{w^3} - \alpha^2 w\right) dw, \tag{10}$$

where $\rho = 0$ when $w = \delta$. Integrating (10), using the boundary conditions, and solving for ρ gives

$$\frac{dw}{du} = \rho = \pm\sqrt{\beta^2\left(\frac{1}{\delta^2} - \frac{1}{w^2}\right) - \alpha^2(w^2 - \delta^2)},$$

which upon using the substitution $q = w^2 - \delta^2$ becomes

$$\frac{\delta}{2\sqrt{\beta^2 q - \alpha^2\delta^2 q(q + \delta^2)}}\, dq = \pm du,$$

where $q = 0$ when $u = 0$. Integrating, using the initial condition, and rewriting in terms of w and u gives

$$\sin^{-1}\left(\frac{2\alpha^2\delta^2}{\beta^2 - \alpha^2\delta^4}\left(w^2 - \frac{\beta^2 + a^2\delta^4}{2\alpha^2\delta^2}\right)\right) = \pm 2\alpha u - \frac{\pi}{2},$$

which in turn simplifies to

$$w^2 = \frac{\beta^2 + \alpha^2\delta^4}{2\alpha^2\delta^2} - \frac{\beta^2 - \alpha^2\delta^4}{2\alpha^2\delta^2}\cos(2\alpha u).$$

Using the double angle formula for cosine, this expression simplifies to

$$w = \sqrt{\delta^2\cos^2(\alpha u) + \left(\frac{\beta}{\alpha\delta}\right)^2\sin^2(\alpha u)}. \tag{11}$$

(We choose the positive square root because $w > 0$ near $u = 0$.) This gives the solutions to (8):

$$\boxed{r(t) = \sqrt{R^2\cos^2(\sqrt{k}\,t) + \frac{h^2}{kR^2}\sin^2(\sqrt{k}\,t)}} \tag{12}$$

Pebble's distance from earth's center with respect to time

with $\delta = R$, $\alpha = \sqrt{k}$ and $\beta = h$, and

$$r(\theta) = \cfrac{1}{\sqrt{\cfrac{1}{R^2}\cos^2\theta + \cfrac{kR^2}{h^2}\sin^2(\theta)}}, \tag{13}$$

Pebble's distance from earth's center with respect to angle

with $\delta = 1/R$, $\alpha = 1$, $\beta = \sqrt{k}/h$ and $r(\theta) = 1/z(\theta)$.

The polar plot of (13),

$$(X(\theta),\ Y(\theta)) = r(\theta)(\cos\theta,\ \sin\theta), \tag{14}$$

is an ellipse of semi-axial lengths R and $h/(R\sqrt{k})$ because $X^2/R^2 + Y^2/(h/(R\sqrt{k}))^2 = 1$. For example, Figure 4a shows the ellipse given by (14) when $h = 0.4R^2\sqrt{k}$.

If we assume that the tangential velocity of pebble P at its moment of being dropped is such that r begins to decrease, then it is clear that the nearest P gets to earth's center is

$$\boxed{\frac{h}{R\sqrt{k}}.} \tag{15}$$

Pebble's perigee

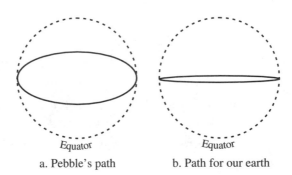

a. Pebble's path   b. Path for our earth

Figure 4. Elliptical solutions in the equatorial plane of the earth

The ellipse of Figure 4b shows the pebble's path for our earth, with $g = 9.8$ m/sec^2, $R = 6400$ km, $k = g/R$ and $Q = 86400$ seconds. Thus the nearest that P approaches earth's center as it falls is a bit more than 376 km.

Now let $A(t)$ be the area swept out by P's position vector from time 0 to time t. Using (3) the change in area ΔA per change in time Δt is given by

$$\Delta A = \int_t^{t+\Delta t} \frac{r^2}{2} \frac{d\theta}{d\tau} d\tau = \int_t^{t+\Delta t} \frac{h}{2} d\tau = \frac{h}{2} \Delta t. \tag{16}$$

That is, P's position vector sweeps out area at a constant rate with respect to the ellipse's center rather than with respect to its focus. This argument is independent of the gravitational field $f(r)$, so Kepler's second law applies to any gravity field radiating from the origin.

A useful form of (12) is the parameterization

$$\boxed{(x(t),\ y(t)) = \left(R\cos(\sqrt{k}\,t),\ \frac{h}{R\sqrt{k}} \sin(\sqrt{k}\,t) \right).} \tag{17}$$

Parametrization of pebble's orbit in the equatorial plane

To see that this parameterization is valid, first observe that both (14) and (17) represent the same ellipse. To demonstrate that they both trace out this ellipse in time in the same way, define the angle $v(t)$ and the length $s(t)$ so that $\tan(v) = y/x$ and $s(t) = \sqrt{x^2 + y^2}$. Therefore

$$\frac{s^2}{x^2} \frac{dv}{dt} = \sec^2 v \frac{dv}{dt} = \frac{xy' - yx'}{x^2}$$

where $x' = \frac{dx}{dt}$ and $y' = \frac{dy}{dt}$. So

$$s^2\, dv = (xy' - yx')\, dt$$

$$= \left(R\cos(\sqrt{k}\,t) \frac{h}{R\sqrt{k}} \sqrt{k} \cos(\sqrt{k}\,t) \right.$$

$$\left. - \frac{h}{R\sqrt{k}} \sin(\sqrt{k}\,t)(-R\sqrt{k}) \sin(\sqrt{k}\,t) \right) dt$$

$$= h\, dt,$$

which means that the area swept out by the position vector (x, y) from t to $t + \Delta t$ is

$$\Delta A = \frac{1}{2} \int_t^{t+\Delta t} s^2 \frac{dv}{d\tau} d\tau = \frac{1}{2} \int_t^{t+\Delta t} h\, d\tau = \frac{h}{2} \Delta t. \tag{18}$$

By (16) and (18), from time 0 to t, the curves in (14) and (17) both sweep out area $ht/2$; and since (17) starts at the initial condition $(R, 0)$ of (5) and proceeds in the counterclockwise direction, the arclengths generated by these two parameterizations over this time period are the same, which means that (17) faithfully describes the motion of the falling pebble.

Is there an analog of Kepler's third law for the period? From (13), we see that the angle lapse between perigee and apogee is $\pi/2$, which by (12) means that the period T of the ellipse is

$$T = \frac{2\pi}{\sqrt{k}}. \tag{19}$$

Period of pebble's orbit in the earth

For earth, this period is about 84.6 minutes, which is the period of a pebble dropped at the north pole through a homogeneously dense earth. If a pebble is dropped initially at distance ρ from the origin in this gravitational field with initial rotation rate about the origin as $2\pi/Q$, then it moves on an ellipse with semi-major axis ρ and semi-minor axis $2\pi\rho^2/(\rho Q\sqrt{k}) = 2\pi\rho/(Q\sqrt{k})$ by (6) and (15); since the ratios of these semi-axial lengths is a constant, then this entire family of ellipses shares the same eccentricity. By (19), the periods of orbits are independent of the initial rotation rate of the pebble about the origin.

To recap, the path of a pebble falling through the earth has the following three properties which the reader may contrast to Kepler's three planetary laws of motion in Chapter IV.

Kepler's three laws where $f(r) = -kr$:   (20)

i. Pebble P's path is an ellipse whose center is the center of the earth. This is corollary i of proposition 10 of book i of the *Principia*.

ii. P's position vector (from the origin) sweeps out area at a constant rate. This is proposition i of the *Principia*, valid for any radial gravity field.

iii. The family of orbits pertaining to pebbles dropped in this field shares the same period. This is corollary ii of proposition 10 of the *Principia*.

If the pebble is dropped from the earth's surface somewhere besides the equator the solution is generated as before, except for the constant of angular momentum. To indicate this new value, let $\hat{h}$ be the angular momentum of the pebble about the earth's center when it is dropped at latitude ψ, where $-\pi/2 \le \psi \le \pi/2$. In Figure 5, the letters A, B, C, D, O and P respectively represent the points $(R, 0, 0)$, $(0, R, 0)$, the north pole, the pebble's drop point $(R \cos \psi, 0, R \sin \psi)$, the center of the earth, and the pebble's position at time t. The gray curve through A and B is the equator; the curve through B and D is a great circle. The tangential speed v of a point at latitude ψ on the surface of the earth about the earth's axis is $2\pi R \cos \psi / Q$, where R is the radius of the earth and Q is the earth's period about its axis; this tangential speed v of the pebble is also $v = R \frac{d\theta}{dt}|_{t=0}$ where $\theta(t)$ is the angle between D and P. Hence $\frac{d\theta}{dt}|_{t=0} = 2\pi \cos \psi / Q$, which means that $\hat{h} = 2\pi R^2 \cos \psi / Q = h \cos \psi$. The pebble's path thus is parametrized by $\left(R \cos(\sqrt{k}\, t), h/(R\sqrt{k}) \cos \psi \sin(\sqrt{k}\, t)\right)$, and lies in the plane through B, D and O. In terms of the standard xyz coordinate system, the pebble's path is obtained by rotating the vectors $(R \cos(\sqrt{k}\, t), h/(R\sqrt{k}) \cos \psi \sin(\sqrt{k}\, t), 0)$ clockwise (with respect to standing at B and looking at O) about the y-axis by ψ radians:

$$
\begin{aligned}
&\big(x(t, \psi), y(t, \psi), z(t, \psi)\big) \\
&= \Big(R \cos \psi \cos(\sqrt{k}\, t), \tfrac{h}{R\sqrt{k}} \cos \psi \sin(\sqrt{k}\, t), R \sin \psi \cos(\sqrt{k}\, t)\Big).
\end{aligned}
\tag{21}
$$

Parametrization of pebble's orbit when dropped at latitude ψ

When $\psi = \pi/2$ the solution path is a straight line between the north and south poles.

The path through the earth

Although we have determined the path of a pebble P falling through the earth with respect to a stationary plane, the earth is rotating. We now change our orientation so that the frame of reference is the rotating earth. We accordingly ask, What is the shape of a hole we must construct in the earth so that a falling pebble never strikes its walls? To answer this question, we find the relationship between θ and t; in particular, we write θ as a function in terms of t, and then graph

$$
r(t) \left(\cos\left(\theta(t) - \frac{2\pi t}{Q}\right), \, \sin\left(\theta(t) - \frac{2\pi t}{Q}\right) \right),
\tag{22}
$$

Parametrization of pebble-hole through the earth

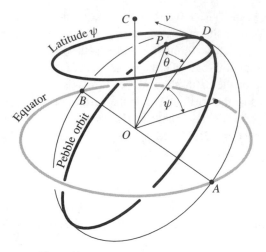

Figure 5. Pebble dropped at latitude ψ

where Q is the period of the earth about its axis and $r(t)$ is as in (12). To derive a relation between θ and t, we start with (3) and (12),

$$\frac{d\theta}{dt} = \frac{h}{R^2 \cos^2(\sqrt{k}t) + \frac{h^2}{kR^2} \sin^2(\sqrt{k}t)} \tag{23}$$

and in the process give an alternate derivation of (17).

To simplify matters let

$$\rho(t) = \frac{1}{a^2 \cos^2(\omega t) + b^2 \sin^2(\omega t)}$$

where a, b, ω are constants, and let $\Theta(t)$ be that function for which $\frac{d\Theta}{dt} = \rho(t)$ with $\Theta(0) = 0$. Since $\Theta(t)$ is simply the area under the always positive, periodic function ρ over the interval from 0 to t, $\Theta(t)$ increases without bound.

Integrating ρ gives

$$\int \rho(t)\,dt = \int \frac{1}{a^2 \cos^2(\omega t) + b^2 \sin^2(\omega t)}\,dt$$

$$= \frac{1}{ab\omega} \tan^{-1}\left(\frac{b}{a}\tan(\omega t)\right) + C, \tag{24}$$

where C is a constant of integration. A disappointing feature of (24) is that $1/(ab\omega) \tan^{-1}(\omega t)$ is bounded above by $\pi/(2ab\omega)$. Rather than just

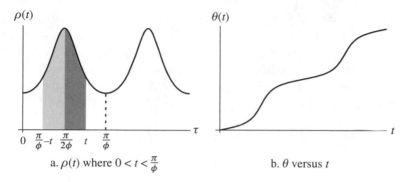

a. $\rho(t)$ where $0 < t < \frac{\pi}{\phi}$ b. θ versus t

Figure 6. The relation between θ and t

applying the antiderivative formula of (24) to compute $\Theta(t)$, we exploit $\rho(t)$'s periodicity. From (24) it is clear that

$$\Theta(t) = \frac{1}{ab\omega} \tan^{-1}\left(\frac{b}{a}\tan(\omega t)\right), \quad \text{for } 0 \le t \le \frac{\pi}{2\omega},$$

and

$$\Theta\left(\frac{\pi}{2\omega}\right) = \frac{\pi}{2ab\omega}.$$

From Figure 6a, for any t with $\pi/(2\omega) < t < \pi/\omega$, the area under the curve from $\pi/(2\omega)$ to t is equal to the area from $\pi/\omega - t$ to $\pi/(2\omega)$. Therefore the area from 0 to t is the area from 0 to π/ω minus the area from 0 to $\pi/\omega - t$, which gives

$$\Theta(t) = \frac{\pi}{ab\omega} - \Theta\left(\frac{\pi}{\omega} - t\right), \quad \text{for } \frac{\pi}{2\omega} < t \le \frac{\pi}{\omega}.$$

For $t > \pi/\omega$, let $n(t) = \lfloor \omega t/\pi \rfloor$ and let $q(t) = t - \pi n(t)/\omega$, the integral number of times that π/ω divides t and the remainder, respectively. Then

$$\Theta(t) = \frac{n(t)\pi}{ab\omega} + \Theta(q(t)) \quad \text{for } t > \frac{\pi}{\omega}.$$

Thus $\theta(t) = h\Theta(t)$ for all t, with $a = R, b = \dfrac{h}{R\sqrt{k}}$, and $\omega = \sqrt{k}$. In particular,

$$\boxed{\theta(t) = \tan^{-1}\left(\frac{h}{R^2\sqrt{k}}\tan(\sqrt{k}t)\right), \quad \text{for } 0 < t < \frac{\pi}{2\sqrt{k}}.} \quad (25)$$

Pebble's angle with respect to time

This means that $\tan\theta = h/(R^2\sqrt{k})\tan(\sqrt{k}\,t) = y/x$, thus corroborating (17). The graph of $\theta(t)$ is shown in Figure 6b. The curve looks like the graph of $at+b\sin(\gamma t)$ for some constants a, b and γ, a bounded, periodic oscillation around a line. With this in mind, consider the expression $\tan\alpha = c\tan\beta$ for values α, β and c. Let $\chi = \alpha - \beta$. By the addition identity for the tangent function,

$$\frac{\tan\beta + \tan\chi}{1 - \tan\beta\tan\chi} = \tan(\beta + \chi) = c\tan\beta.$$

Solving for $\tan\chi$ gives

$$\tan\chi = \frac{(c-1)\sin\beta\cos\beta}{1 + (c-1)\sin^2\beta},$$

which means that

$$\alpha = \beta + \tan^{-1}\left(\frac{(c-1)\sin\beta\cos\beta}{1 + (c-1)\sin^2\beta}\right). \tag{26}$$

Substituting $\alpha = \theta$, $\beta = \sqrt{k}\,t$, and $c = h/(R^2\sqrt{k})$ in (26) gives an improvement over (25) for $\theta(t)$, although it is not quite of form $at + b\sin(\gamma t)$:

$$\boxed{\begin{aligned}&\theta(t)\\ &= \sqrt{k}\,t + \tan^{-1}\left(\frac{\left(\frac{h}{R^2\sqrt{k}}-1\right)\sin(\sqrt{k}\,t)\cos(\sqrt{k}\,t)}{1+\left(\frac{h}{R^2\sqrt{k}}-1\right)\sin^2(\sqrt{k}\,t)}\right) \text{ for } t \geq 0.\end{aligned}} \tag{27}$$

Pebble's angle with respect to time, alternate version

The inverse tangent term of (26) is bounded and periodic because its argument is periodic, continuous and has positive denominator. The path that the pebble drills through the earth (neglecting friction) is in Figure 7, as given by (22).

For earth, a pebble dropped at the equator will return to its drop point with respect to a nonrotating coordinate system in 84.6 minutes. Because the earth rotates, the geographical point where the pebble returns is actually about 2360 km due west of the geographical drop point, as illustrated in Figure 7.

In a lengthy discussion [35, p. 139], Galileo examined the experiment of dropping a cannonball from a lofty tower. Data of that day confirm that the ball falls straight down, seeming evidence that the earth fails to rotate. To illustrate, let's use the leaning Tower of Pisa in northern Italy, 55 meters tall,

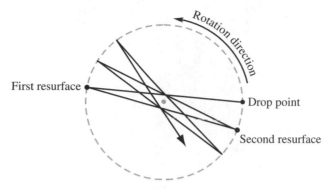

Figure 7. The hole through the earth

atop which Galileo is said to have performed his legendary experiment of
dropping cannonballs of disparate masses, so confuting Aristotle's premise
that more massive objects fall faster than less massive objects. The tower
in Figure 8 is from a painting, viewable at [**30**]. As the figure shows, the
top floor of the Tower is recessed. So let's drop a ball from 50 meters; its
flight time (which Galileo could calculate) is about 3.2 seconds. From such
a height, and at Pisa's latitude of $\psi = 44°$, a Ptolemaic advocate might
reason, "If the earth rotates once per day, then the ball should fall west of
the tower by about $(6400 \text{ km})(\cos \psi)(2\pi/Q)(3 \text{ seconds}) \approx 1$ km; but since
no westward displacement occurs, then the earth is fixed in space." Against
such logic, a Copernican advocate would reason, "Just as a ball dropped from
atop a tall mast on a moving ship falls straight down, so too on the earth,"
a rationale anticipating the notion of a Newtonian reference frame. But a
ball dropped from the tower actually falls eastward, the reason being that
the tangential velocity of the top of the tower on a rotating earth is greater
than the tangential velocity at its base, so as the ball falls it continues to go
eastward slightly faster than the base of the tower, and thus strikes the ground
a bit to the east of straight down. Figure 7 illustrates this phenomenon: draw
a line from the drop point to the earth's center—the hole lies to the east of
this line.

In particular, if a pebble falls from the surface at the equator, then at time t
the pebble is at $r(t) \, (\cos(\theta(t) - 2\pi t/Q), \sin(\theta(t) - 2\pi t/Q))$ with respect
to the rotating equatorial plane. Measured along the circumference of a circle
of radius $r(t)$ and center $(0, 0)$, the eastward displacement of the pebble
away from a vertical line between the earth's center and the geographical
drop point is the arclength $r(t) \, (\theta(t) - 2\pi t/Q)$. By (12) and either (25) or

Figure 8. Tower at Pisa (courtesy of G. Feuerstein [30])

(27), this displacement $E(t)$ is

$$E(t) = \left(\tan^{-1}\left(\frac{h}{R^2\sqrt{k}}\tan(\sqrt{k}t)\right) - \frac{2\pi}{Q}t\right)$$
$$\times \sqrt{R^2\cos^2(\sqrt{k}t) + \frac{h^2}{kR^2}\sin^2(\sqrt{k}t)}, \tag{28}$$

Eastward displacement of pebble in the equatorial plane

for $0 < t < \pi(2\sqrt{k})$. The reader may show using (21) that the eastward displacement $E(t, \psi)$ of the falling pebble when dropped at latitude ψ is

$$E(t, \psi) = E(t)\cos\psi. \tag{29}$$

Eastward displacement of pebble when dropped at latitude ψ

Under the assumption of no air resistance, a little calculation using (28) shows that a drop down a hole of 50 meters at the equator results in the pebble landing about 8 mm east of straight down. The same result occurs if the pebble is dropped from a tower of 50 meters rather than down a hole. At

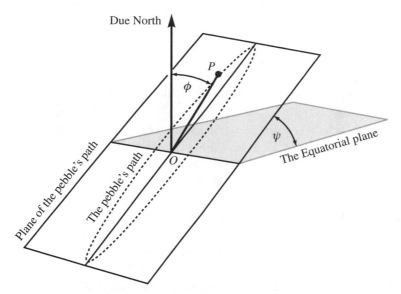

Figure 9. Finding the southward displacement

latitude $44°$, the same drop gives a displacement of about 6 mm from (29), not much of an eastward displacement for Galileo's cannonball.

A pebble dropped at the surface of the earth at some place other than the equator falls toward the equator as well as to the east. To verify this, let's drop a pebble at latitude ψ in the northern hemisphere so that the drop point is $(R \cos \psi, 0, R \sin \psi)$. Let $\phi(t, \psi)$ be the angle between due north and the pebble's position at time t, point P on Figure 9. From (21), the pebble's distance from the center of the earth is

$$r(t, \psi) = \sqrt{R^2 \cos^2(\sqrt{k}\, t) + \frac{h^2}{kR^2} \cos^2 \psi \sin^2(\sqrt{k}\, t)}. \tag{30}$$

Thus by (21) and (30),

$$\phi(t, \psi) = \cos^{-1}\left(\frac{z(t, \psi)}{r(t, \psi)}\right)$$

$$= \cos^{-1}\left(\frac{R \sin \psi \cos(\sqrt{k}\, t)}{\sqrt{R^2 \cos^2(\sqrt{k}\, t) + \frac{h^2}{kR^2} \cos^2 \psi \sin^2(\sqrt{k}\, t)}}\right). \tag{31}$$

While the pebble falls, the point we dropped it from remains at latitude ψ. Measured along the surface of a sphere centered at the origin and of radius $r(t, \psi)$, the pebble's southward displacement away from this sphere's circle of latitude ψ is

$$S(t, \psi) = \left(\phi(t, \psi) - \left(\frac{\pi}{2} - \psi\right)\right) r(t, \psi). \qquad (32)$$

Southward displacement of pebble when dropped at latitude ψ

Evaluating (32) for Galileo's ball, we find that the ball should fall about 8.7 cm south of straight down. So Galileo was ironically close to having conclusive evidence that the earth rotates! However, measuring this displacement is challenging even if you know to look for it. Although Galileo did not look for eastward or southward displacements, others soon did.

In December of 1679, following Newton's suggestion, Hooke, who as curator of the Royal Society regularly performed experiments for its weekly meetings, repeated the experiment, this time in a cathedral with its windows and doors closed so as to minimize air currents. Dropping a ball from 9 meters, Hooke measured the impact as at least a quarter of an inch southeast of what he thought was straight down [6, pp. 20–1]. By (29) and (32), a fall from 9 meters at latitude 52° gives an eastward displacement of 0.36 mm, and a southward displacement of 1.5 cm $\approx$ 0.59 inches, values that at first glance seem to confirm Hooke's result.

In 1831, F. Reich dropped pellets 188 meters down a mine shaft, measuring the displacements as about 2.8 cm southeast of what he thought was straight down [52, p. 395]. If his mine shaft was at latitude 45° with a drop time of 6.2 seconds, (29) and (32) predict an eastward displacement of 4 cm and a southward displacement of 32 cm, predictions which seem far afield, especially the southward displacement. What's going on?

To resolve the disparity between these experiments and the corresponding theoretical southern displacements, define straight down as the ray from the drop site to the center of the earth, and define apparent down as the ray along a plumb line from the top of the plumb line to the equilibrium point at its lower extremity. Except at the poles and at the equator, straight down and apparent down are not the same direction. At 45°, the discrepancy between them is about 0.1°. An exercise in [52, p. 403, exercise 14] shows that for objects dropped 200 meters (no resistance) at the earth's surface, the southward displacement of the object away from apparent down is about 0.01 mm. So southerly displacements (using a plumb bob) for short drops are basically unmeasurable. As the technology in global positioning satellites improves, the common notion of down may eventually change to mean straight down; if it does, then finding the southerly displacement of a dropped object will be a surprising introductory physics experiment.

Someday, someone might try the experiment on the moon, exulting in the turbulence-free conditions. In anticipation of this event, here are the vital statistics for the moon: $g = 1.57$ m/s^2, $R = 1741$ km, $f(r) = -kr$, $k = g/R$, $Q = 2.36$ million seconds. At latitude 45°, dropping a pebble down a deep lunar crevasse of 1 km gives a fall time of 36 seconds, a 4.5 cm eastward displacement and a 4 mm southward displacement from straight down. That is, plumb bobs on the moon point almost straight down.

Comparing the eastern and southern displacements

In this section we take a closer look at the east-south displacement of a falling pebble, which is most pronounced at latitude $\psi = \pi/4$ in the northern hemisphere. Figure 10 gives the eastward and southward displacements of a pebble falling from earth's surface at 45°N, and is the comparison of the graphs of $E(t, \pi/4)$ and $S(t, \pi/4)$ of (29) and (32). The eastward displacement overtakes the southward displacement after about 50 seconds of fall. During this interval the pebble remains so near earth's surface that it does not matter whether the pebble is falling down a hole or from a balloon overhead.

To convey the dynamics of Figure 10 better, we normalize the southward and eastward displacement by dividing by the vertical distance drop in time t, the distance $R - r(t, \psi)$, which is approximately $\frac{1}{2}gt^2$. We define the normalized southward displacement of a pebble dropped at (radian) latitude ψ by $\hat{S}(t, \psi) = 100S(t, \psi)/(R - r(t, \psi))$ and the normalized eastward displacement by $\hat{E}(t, \psi) = 100E(t, \psi)/(R - r(t, \psi))$. Thus $\hat{S}$ and $\hat{E}$

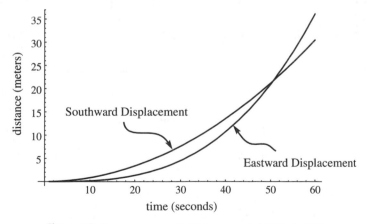

Figure 10. East versus south displacement at latitude 45°

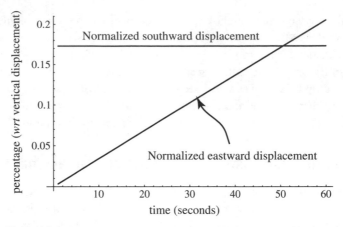

Figure 11. Normalized east versus south displacement at latitude 45°

measure the southward and eastward displacements with respect to the vertical drop of the pebble. Figure 11 shows them at 45°N latitude. The normalized southward displacement is almost a constant 0.1%, and it remains virtually unchanged all the way to the pebble's closest approach to earth's center. On the other hand, the normalized eastward displacement appears to be linear, and continues in that way as it falls towards its perigee. That is, it appears as if the first non-zero term in a Taylor series expansion for $S(t, \pi/4)$ is of degree 2, whereas for $E(t, \pi/4)$ it is of degree 3.

To check this conjecture, we use the following first-order approximations:

$$\sin \theta \approx \theta, \quad \text{when } \theta \text{ is near } 0. \tag{33}$$

$$\cos \theta \approx 1, \quad \text{when } \theta \text{ is near } 0. \tag{34}$$

$$\cos^{-1} w \approx \frac{\pi}{4} - \sqrt{2}\left(w - \frac{1}{\sqrt{2}}\right), \quad \text{when } w \text{ is near } 1/\sqrt{2}. \tag{35}$$

$$\frac{1}{\sqrt{1+z}} \approx 1 - \frac{z}{2}, \quad \text{when } z \text{ is near } 0. \tag{36}$$

We derive an approximate formula for $S(t, \pi/4)$ for small t. From (31) and with $\psi = \frac{\pi}{4}$,

$$\cos \phi = \frac{R \cos(\sqrt{k}t)}{\sqrt{2R^2 \cos^2(\sqrt{k}t) + \frac{h^2}{kR^2} \sin^2(\sqrt{k}t)}}.$$

By (33) and (34) with $\theta = \sqrt{k}t$,

$$\cos \phi \approx \frac{R}{\sqrt{2R^2 + \frac{h^2 t^2}{R^2}}}.$$

By (6),

$$\cos \phi \approx \frac{1}{\sqrt{2}} \frac{1}{\sqrt{1 + \left(\frac{\sqrt{2}\pi t}{Q}\right)^2}}.$$

By (36) with $z = (\sqrt{2}\pi t/Q)^2$,

$$\cos \phi \approx 1/\sqrt{2} - 1/\sqrt{2}(\pi t/Q)^2.$$

By (35) with $w = 1/\sqrt{2} - (1/\sqrt{2})(\pi t/Q)^2$,

$$\phi \approx \frac{\pi}{4} - \sqrt{2}\left(-\frac{1}{\sqrt{2}}\left(\frac{\pi t}{Q}\right)^2\right) = \frac{\pi}{4} + \left(\frac{\pi t}{Q}\right)^2.$$

So $S(t, \pi 4) \approx (\pi t/Q)^2 R$, a second-order term as expected, for small t. It similarly follows that for small t,

$$\boxed{S(t, \psi) \approx \left(\frac{\pi t}{Q}\right)^2 R \sin(2\psi).}$$

A Taylor approximation for southward displacement of falling pebble

Furthermore, using a third-order and a first-order Taylor approximation for the tangent and inverse tangent functions respectively, the reader may similarly derive an approximation for $E(t, \psi)$ for small t,

$$\boxed{E(t, \psi) \approx \frac{2\pi g t^3 \cos \psi}{3Q}.} \tag{37}$$

A Taylor approximation for eastward displacement of falling pebble

That is, as given by Figure 11, our intuition as to the structure of $E(t, \psi)$ and $S(t, \psi)$ was correct.

To help understand the significance of this east-south relationship, imagine being in a hot-air balloon on a still day at latitude 45°N. A prize is given to whomever can drop a sandbag nearest to a bullseye 1100 meters below the balloon using eyesight alone. Assume that the balloon has a propeller so that it can be maneuvered conveniently. From 1100 meters the sandbag strikes the ground in about 15 seconds. Figure 10 shows that we must aim much more north than west, about two meters north and about half a meter west, of the bullseye to strike dead center.

General gravity fields

Up to now, we have modeled the pebble's falling motion with respect to a linear acceleration function. But there is no need to be restrictive. In fact, earth's actual acceleration function is nonlinear, as is pointed out in Chapter II. Newton began the *Principia* by considering a pebble falling through the earth, primarily because an inverse square acceleration is harder to deal with (than a linear one). However, he didn't belabor the details of the linear acceleration case because it was academic. After all, a pebble can't drill its own hole through the earth. Or can it?

Consider a meteor plunging through a nebula, an astronomical object that can be thought of as a nebulous planet; as such the meteor is plunging through a planet whose acceleration function is neither linear nor an inverse square. We can also consider a star falling through a galaxy, where we think of the star as a pebble and the galaxy as a planet. So, the problem of this chapter passes from being an academic curiosity to a potential model for interesting astronomical phenomena. For that reason, we generalize the falling pebble problem to an arbitrary acceleration function. In doing so, we first characterize the structure of these gravity field functions, and then generate paths of pebbles falling through such fields.

The standard linear gravitational acceleration function field

$$
f_L(r) = \begin{cases} -\dfrac{g}{R}r, & 0 \le r \le R, \\[2ex] -\dfrac{gR^2}{r^2}, & r > R, \end{cases}
$$

as depicted in Figure 12a, is but one of various models for the earth's gravity, where r is distance from the center of the earth, $g = 9.8$ m/sec^2, and $R = 6400$ km, the radius of the earth. According to geophysical research, the earth's gravity at the juncture between its outer core and its mantle, 3500 km from the center, is about -10.8 m/sec^2. Figure 12b is a fairly accurate approximation f_M of earth's actual gravity field; see [51, p, 155] for today's current best guess for earth's gravity field. Up to the earth's radius, f_M is piecewise linear, connecting the data 0, -10.8, and -9.8 at the center, the core-mantle juncture, and the earth's surface, respectively; thereafter, f_M

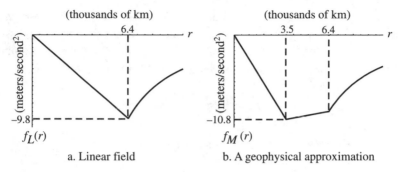

a. Linear field   b. A geophysical approximation

Figure 12. Gravity fields for the earth

follows the inverse square law. Numerically solving (4) using f_M shows that the pebble comes within 302 km of the earth's center in about 19.1 minutes, which beats f_L's values of 376 km and 21.2 minutes. With this problem in mind, it is natural to solve (4) generally for nonlinear gravity fields $f(r)$. In this section, we formalize what is meant by a gravity field, explore several fields, make some observations and definitions, and ask some general questions.

Let Ω be a planet whose density is spherically symmetric, and let $\delta(r)$ be the density of Ω at radial distance r. We specify that $\delta(r)$ be a non-negative, real-valued, piecewise continuous function on the interval $0 < r \leq R$, where R is the radius of Ω, and that δ is identically 0 when $r > R$. From (7) in Chapter II, the gravitational acceleration, or gravity field, f for Ω is

$$f(r) = -\frac{4\pi G}{r^2} \int_0^r \rho^2 \delta(\rho) \, d\rho, \text{ for } r > 0, \qquad (38)$$

Gravity field for a planet in terms of its density

where G is the universal gravitational constant and ρ is a dummy variable for radial distance. For any $r > R$, (38) implies that $f(r) = -f(R)R^2/r^2$. Since we wish to solve (4) for general gravity fields, it would be convenient to have a simple test for detecting them. With this in mind, we observe that any function $f(r)$ which has hopes of being a gravity field must have the following properties:

i. $f(r) \leq 0$ for all $r \geq 0$.

ii. f is continuous on the interval $[0, \infty)$.

iii. f is piecewise differentiable on the interval $(0, \infty)$.

iv. For each $a \geq 0$ with $f(a) < 0$, $f(r) < 0$ for all $r \geq a$.

Any function f that has these four properties we call taut. (We use *taut* because of property (iv) that the gravitational effect never completely dissipates.) To find the taut functions that are gravity fields, we use (38) to find the density function $\delta(r)$ for a given gravity field. To do so, rewrite (38) as

$$r^2 f(r) = -4\pi G \int_0^s \rho^2 \delta(\rho) \, d\rho$$

and differentiate with respect to r, obtaining $r^2 f'(r) + 2rf(r) = -4\pi Gr^2 \delta(r)$, which when solved for $\delta(r)$ gives

$$\delta(r) = -\frac{2f(r) + rf'(r)}{4\pi Gr}. \tag{39}$$

Density for a planet in terms of its gravity field

This means that for any gravity field $f(r)$, $rf'(r) + 2f(r)$ is non-positive whenever f' is defined. Since every gravity field $f(r)$ is non-positive, this means that provided $f(r)$ is non-zero and $f'(r)$ is defined,

$$-\frac{f'(r)}{f(r)} \leq \frac{2}{r}. \tag{40}$$

A characterization for gravity fields

If f is taut and (40) holds, these steps are reversible. In particular, given that f is negative, (40) means that $-f'(r) \geq 2f(r)/r$, which in turn implies that $2f(r) + rf'(r) \leq 0$, which shows that the right-hand side of (39) is non-negative, making it a density function, and so f as given by (38) is a gravity field. Thus (40) is a test for detecting which taut functions are gravity fields.

For any $a > 0$, we say that f is a gravity field on the interval $[0, a]$ if f is taut and (40) is satisfied on the interval $[0, a]$. For example $f(s) = -e^{-s}$ fails to be a gravity field on $[0, a]$ for sufficiently large a. To see this, for this function (40) becomes $1 \leq 2/s$, which means that this function behaves as a gravity field on $[0, a]$ only when $0 \leq a \leq 2$.

As another example, let $f_1(s) = -2h^2 s^2$ and $f_2(s) = h^2 s(s - 3)$, where h is the constant given by (6). Since both f_1 and f_2 are non-positive, differentiable, and decreasing on the interval $[0, 1]$, (40) is trivially satisfied on the interval, which means that both functions behave as gravity fields on the unit interval.

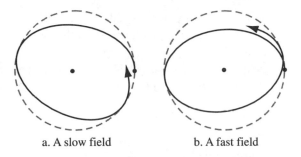

a. A slow field   b. A fast field

Figure 13. Precessing ellipses

Two fields f and g on $[0, R]$ are said to be gravity fields modeling a planet of radius R if $f(R) = g(R)$. When we refer to planet Ω and make reference to its gravity field f we mean the one inherently given by its density. We say that a pebble-path for planet Ω is the path generated by (4) and (5) under its inherent gravity field and its rotation rate. The apogee and perigee of a pebble-path are the greatest and least radial distances, respectively, of the pebble-path from the planet's center. The apogee and perigee associated with a field are the apogee and perigee of the pebble-path associated with the planet. A proper field for Ω is one wherein Ω's pebble-path has apogee R, Ω's radius. A pebble-hole for Ω is Ω's pebble-path adjusted to Ω's natural rotating coordinate system.

For example, for f_1 and f_2, (7) gives

$$\frac{d^2z}{d\theta^2} + z = \frac{2}{z^4} \quad \text{and} \quad \frac{d^2z}{d\theta^2} + z = \frac{3z-1}{z^4},$$

respectively. For planets of radius 1, the pebble-paths associated with these two gravity fields are shown in Figure 13.

To discuss the shapes of these paths with greater facility, we stipulate some additional constraints and definitions. Planet Ω spins at a positive rate counterclockwise viewed from above the polar plane. The standard field for Ω is $f_0(r) = -kr$, where k is a positive constant. The speed of any proper field for Ω is $1/\pi$ times the angle measured counterclockwise between successive apogee occurences for the pebble-path. For example f_0's speed is 1, whereas the speed of f_1 is less than 1 and the speed of f_2 is more than 1. Let the support of a function be the set of all points in the interval $[0, \infty)$ where the function is nonzero. A function is said to have positive support if its support contains an open interval. If f and f_0 model the same planet, field f is fast if $f(r) \le f_0(r)$ for all r between 0 and radius R and if $f - f_0$ has positive support. Similarly, a field f is slow if this inequality is reversed. Note that

f_1 is slow and f_2 is fast for a planet of radius 1. For two fields f and g modeling a planet of radius R, f is said to be faster than g if $f(r) - g(r)$ is non-positive and has positive support.

We close this chapter with exercises whose solutions demonstrate that the hole drilled by a falling pebble through its planet can be exotic.

Exercises

1. **Well-definedness.** With respect to any planet, prove that every proper field has a well-defined speed. That is, if a pebble is dropped, then it returns to the surface.

2. **Speed comparison.** With $R = 1$, find the speeds of f_1 and f_2. Prove or disprove: a fast proper field's speed exceeds 1, whereas a slow proper field's speed is less than 1.

3. **Perigee comparison.** Let f and g both model a planet of radius R. Are either of the following intuitive statements true? If f is a faster field than g, then the perigee associated with f is less than the perigee associated with g. Furthermore, the speed of f exceeds that of g. (For example, the piecewise linear gravity field f_M is faster than the standard f_L. The speed of field f_M is about 1.012 whereas the speed of f_L is 1. f_M's perigee is about 302 km whereas f_L's perigee is about 376 km.)

4. **The fastest field.** Prove that for any planet the greatest speed of any proper field is 2 and occurs when $f(r) = -k/r^2$; this field also generates the least perigee.

5. **Proportionality of speed and perigee.** For any planet Ω and real number q with $0 < q \leq R$, let Ω_q be the planet of radius q which has the same density function and rotation rate as Ω. For the standard field, by Kepler's modified third law (20), no matter where the pebble is initially dropped at positive radial length between 0 and R, the speed remains fixed; that is, the speed of the pebble-path for Ω_q is the same for all q with $0 < q \leq R$. Is there a field so that its speed with respect to Ω_q is directly proportional to q?

6. **An analytical solution.** The graphs in Figure 13 are like precessing ellipses. The polar parameterizations

$$r(\theta) = \frac{1}{\sqrt{\cos^2(\alpha\theta) + b^2 \sin^2(\alpha\theta)}},$$

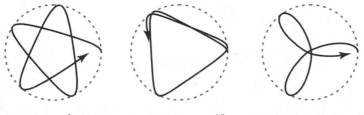

a. $f_a(r) = -7r^3(e^r - 1)$ b. $f_b(r) = -11r^{10}$ c. $f_c(r) = 0.0003r(r-3)^{15}$

Figure 14. A gallery of trochoid-like paths

where b and α are positive constants with $R = 1$, generate curves resembling pebble-paths associated with fast fields when $\alpha < 1$ and slow fields if $\alpha > 1$. For $\alpha \neq 1$, find a variation of this parameterization and a field so that the variation solves (4) for that field. Perhaps begin by looking at the family of fields $f(r) = -kr^n$ where n is a real number.

7. **Through the antipode.** Find a proper field for the earth such that a pebble dropped on the equator at geographical location X will next surface at X's antipode. The standard field does not have this property, as Figure 7 demonstrates.

8. **Exotic paths.** Find fields that yield pebble-paths more exotic than the ones shown in Figure 14 with $R = 1$ and $h = 1$, which are like trochoids. Find a fast field, other than $f(r) = -k/r^2$, whose pebble-hole most resembles the polar flower $r = \cos(3\theta)$. Also, for f_c, find the velocity at which one should shoot a cannonball due west at Ω's surface so that the pebble-path looks like $r = \cos(3\theta)$.

9. **Small-order approximations for south and east abberations.** Show that a second-order approximation for (32) is

$$S(t, \psi) \approx \left(\frac{\pi t}{Q}\right)^2 R \sin(2\psi)$$

and that a third-order approximation for (29) is

$$E(t, \psi) \approx \frac{2\pi g t^3 \cos\psi}{3Q}.$$

10. **Nebula cloud patterns.** The Spirograph Nebula, 2000 light years away, is a low-density, ellipsoidal cloud of matter encircling a small star; it appears to have structure similar to that of the precessing ellipse of Figure 15a. (Permission to use the image of Figure 15b was given by

a. A 2-D spirograph

b. The Spirograph Nebula

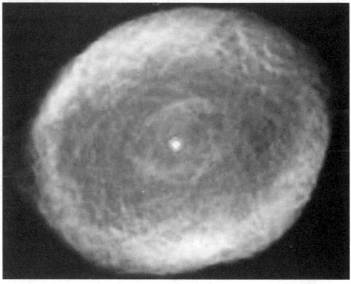

Figure 15. Spirographs

R. Sahai of the Hubble Heritage Team of the Jet Propulsion Laboratory of NASA; a color photo is available on the web [**69**].) In this cloud, each bit of matter is drifting in the gravitational field induced by the matter of the cloud as a whole. Therefore the gravity field in the outer reaches of the nebula little resembles an inverse square law. Each tiny chunk of matter is like a pebble falling through the nebula. Perhaps resistance to motion is negligible in much of the cloud, in which case the paths or currents of matter should trace out spirographic swirls. Perhaps there are larger chunks of matter falling through the nebula that behave as bulldozers or vacuum cleaners sweeping out channels in the clouds, somewhat like jets leaving a trail in the sky or ocean liners leaving enormously long wakes in the sea. Is this gravitational phenomenon part of what helps to create the spirographic structure as captured by the Hubble Telescope in Figure 15b, or is it just coincidence?

preamble ix

Whence this inward horror
Of falling into naught?

Joseph Addison (1672–1719), *Cato*

The sun shone through the office window on a Saturday afternoon. I was grading exams. To avoid getting a headache from marking the same mistakes and writing the same corrective comments over and over, I had music playing low. Gregorian chants work well for me. Since they're mostly in Latin, my mind doesn't get distracted with the words. The pop vocalist, Enya, works well too. I'm not sure what language she sings.

The building shook. The walls trembled, moving at least a quarter inch. Moments later, the building returned to its usual dormant state.

No, this wasn't California. It was east Tennessee. How could an earthquake occur here?

The epicenter was at New Madrid, a little town on the Mississippi where Kentucky, Missouri, and Tennessee meet. Each year, about 150 quakes occur here, with five percent of them being strong enough to feel. The New Madrid Seismic Zone has been active for a long time. Except for the Alaska earthquake of 1964, the earthquake of 1812 at New Madrid is the strongest known quake to have occured in the United States. It is said that an island of river pirates completely disappeared in that quake, and that the Mississippi ran backwards for more than an hour.

The earth is not a homogeneous mass. Its pieces move and change shape under enormous pressure. Fortunately, its surface is stable most of the time. We often take that stability for granted. That's an illusion.

Despite the danger, there's not much that can be done to avoid a natural disaster. The next big one could come in a moment or in a hundred years.

I returned to the stack of papers. It was a nice day. I finished grading them outside, under a tree.

CHAPTER IX

Shadow Lands

As we saw in Chapter II, the path that a body follows when falling through
the earth depends on the earth's structure. How can we know that structure?
A good approach is to use a rule personified by Lewis Carroll's White Rabbit,
who pulls out a pocket watch, looks at it, and then rushes down a hole after
muttering,

Oh dear! Oh dear! I shall be too late!

We shall see that shockwaves passing through the earth are like the White
Rabbit, ever in a hurry. They proceed along paths as if wishing to minimize
their travel time. Observing them can tell us much about the structure through
which they pass. In particular, when an earthquake occurs at point A near the
surface of the earth, not every site is jostled by the event. The very structure
of the earth shields some sites, creating a shadow into which shock waves
from A cannot enter. In this chapter, we derive Snell's law for a flat earth

Figure 1. The White Rabbit in a hurry, illustration by John Tenniel [**18**, p. 12]

141

of horizontal layers, adapt it to a round earth of concentric layers, and use the results along with some data to approximate the depth of the core-mantle boundary.

An intuitive derivation of Snell's law

In 1621, Willebrord Snell discovered that for a light wave,

$$\frac{\sin\theta}{a} = \frac{\sin\phi}{b}, \tag{1}$$

where a is the relative speed of a wave in the region above the x-axis, b is the relative speed in the region below the x-axis, and θ and ϕ are the angles of inclination of the wave away from a normal to the x-axis, as shown in Figure 2.

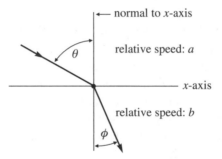

Figure 2. Snell's law

Since it is the primary equation in the chapter, let us derive (1) by using an intuitive model, one that is applicable whether the wave is light, sound, or a shock wave.

In 1657, René Descartes suggested the first intuitive model for light re-fraction. He likened a wave crossing the boundary between two mediums to a ball rolling across a floor from hard-wood onto a rug [**42**, p. 11]. Since an ideal ball has just a single contact point with the floor whereas a wave should have some breadth contact with its medium, let us modify Descartes' ball into the next most simple model, and use a two-point contact with the floor. That is, we use a barbell: two disks $\hat{A}$ and $\hat{B}$ connected by a bar of length d. Each disk makes contact with the floor at a single point, and each traces a path as the barbell rolls along. As the barbell rolls, we stipulate that the disks never slip. What this condition means is that the plane of each disk meets

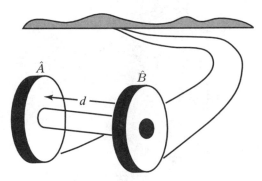

Figure 3. Descartes' barbell

the floor along the tangent line to the path made by the disk. This model, as shown in Figure 3, is enough to derive Snell's law.

As the barbell approaches a straight-line boundary between two mediums, as shown in Figure 4, once disk $\hat{A}$ crosses the boundary, it begins to roll at a different speed than $\hat{B}$. This means that the barbell starts turning, and continues turning until $\hat{B}$ also crosses the boundary. In particular, if $a > b$, then $\hat{B}$ moves further than $\hat{A}$ during the turn; so after the turn, the new angle of inclination of the barbell with respect to the boundary is less than θ.

Except for a change in scale, the bending of the barbell's path as it crosses the boundary is invariant with respect to d. That is, for each d, we choose a unit length equal to the length of the barbell, and a unit time so that the relative speeds of a and b are the speed values of the barbell in the two mediums. Thus we may take d as 1, and interpret it as any distance, including the case when the two disks are arbitrarily close together, which gives a way of approaching the behavior of Descartes' rolling ball model.

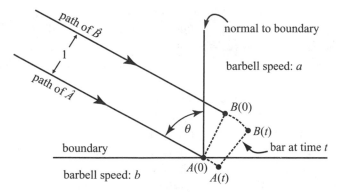

Figure 4. Barbell approaching a boundary

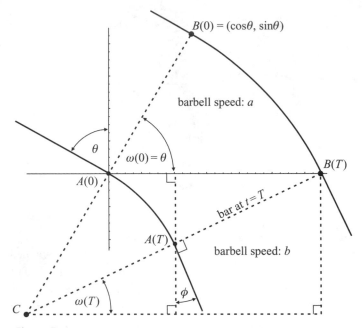

Figure 5. Barbell crossing boundary, $\theta = \pi/3$, $a = 2$, $b = 1$, $T \approx 0.6$

We can find the paths of the barbell disks as they cross the boundary without solving any differential equations. Here's how. Let $A(t) = (x(t), y(t))$ and $B(t) = (u(t), v(t))$ be the positions of $\hat{A}$ and $\hat{B}$ at time t. Let $A(0) = (0, 0)$ and $B(0)$ be above the x-axis. We assume that the paths traced by $\hat{A}$ and $\hat{B}$ are unique. One pair of paths that satisfies the condition that the disks never slip is when they are circular arcs with the same center C, which, by uniqueness, means that they are the paths of the barbell disks.

In particular, C lies along the line through $A(0)$ and $B(0)$. If $a > b$ the center lies below the x-axis, whereas if $b > a$ then the center lies above the x-axis. Let us consider the case when $a > b$. The argument for $a < b$ is very similar.

Let r be the radius of the smaller circle. Thus $1 + r$ is the radius of the larger circle. Let T be the time at which $\hat{B}$ is on the x-axis. Since the smaller sector, with arc from $A(0)$ to $A(T)$, is similar to the larger sector, with arc from $B(0)$ to $B(T)$, then $r/b = (r + 1)/a$, which means that $r = b/(a - b)$ and $C = -b/(a - b)(\cos\theta, \sin\theta)$, as shown in Figure 5. The large radius is $1 + r = a/(a - b)$. At time t, let $\omega(t)$ be the angle between $B(t) - C$ and the horizontal. Thus $\omega(0) = \theta$. Therefore,

$$A(t) = (x(t),\ y(t))$$

$$= -\frac{b}{a-b}(\cos\theta,\ \sin\theta) + \frac{b}{a-b}(\cos\omega(t),\ \sin\omega(t)), \qquad (2)$$

and

$$B(t) = (u(t),\ v(t))$$

$$= -\frac{b}{a-b}(\cos\theta,\ \sin\theta) + \frac{a}{a-b}(\cos\omega(t),\ \sin\omega(t)). \qquad (3)$$

Since $\hat{B}$ is on the x-axis at time T, then $a/(a-b)\sin\omega(T) = b/(a-b)\sin\theta$ by (3), so that

$$\sin\omega(T) = \frac{b}{a}\sin\theta. \qquad (4)$$

By (2) and (4),

$$-y(T) = \frac{b}{a-b}\sin\theta - \frac{b}{a-b}\sin\omega(T) = \frac{b}{a-b}\left(\sin\theta - \frac{b}{a}\sin\theta\right) = \frac{b}{a}\sin\theta.$$

For $t > T$, the barbell proceeds along straight line paths. Let ϕ be the angle of inclination of these paths measured from a vertical to the x-axis, as shown in Figure 5. Then $\omega(T) = \phi$. By alternate interior angles, $\angle A(T)B(T)A(0) = \phi$, which means that $\sin\phi = -y(T)$. Thus

$$\frac{\sin\phi}{b} = \frac{-y(T)}{b} = \frac{b\sin\theta}{ab} = \frac{\sin\theta}{a},$$

which is (1), since the unit length of the bar in the barbell is an arbitrary distance, and can be taken as negligible.

For a contrasting (non-elementary) derivation of Snell's law in the context of Maxwell's equations, see Born [**13**, pp. 36–8].

Through an earth of parallel slabs

In this section, we generalize (1) to an arbitary number of parallel regions, and show that it implies Pierre Fermat's principle that a wave takes the path of least time. That is, we show that shock waves behave like the White Rabbit in arriving as soon as is possible. We then transform (1) into a differential equation.

When an earthquake or explosion occurs, a wave front of disturbance proceeds outward radially in all directions from the event in push-pull surges called longitudinal waves or in side-to-side wriggles called shear waves. Since longitudinal waves are transmitted by both solids and liquids whereas shear waves are transmitted only through solids, we simplify matters and focus on longitudinal waves in this chapter. We can think of a shock wave front as comprised of many radiating strands which we call signals. These signals behave much like light rays, even though any signal probably exists only by being part of a wave front of neighboring signals. We also assume that signals remain undamped as they pass through the earth, a condition like neglecting air resistance for falling bodies.

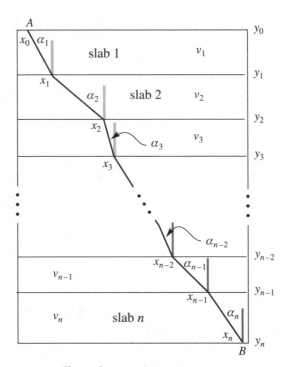

Figure 6. A wave through n layers

Since the earth consists of many dissimilar layers, let us first consider an earth of n parallel layers or slabs, each of which is homogeneous. Let us consider a signal traveling from A to B as in Figure 6 wherein the signal passes through n slabs of material, ever going towards the right and ever descending. In slab i, $1 \le i \le n$, assume that the signal travels with speed v_i

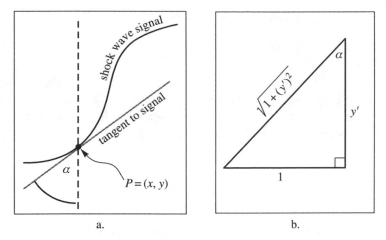

Figure 7. A path through the earth

along line segment $\mathcal{L}_i$ from position (x_{i-1}, y_{i-1}) to position (x_i, y_i). Let α_i be the angle of incidence that $\mathcal{L}_i$ makes with the horizontal boundary lines of slab i. The time lapse T for the signal to travel from A to B (where B is to the right of A) along the n line segments is

$$T = \sum_{i=1}^{n} \frac{\sqrt{(x_i - x_{i-1})^2 + (y_i - y_{i-1})^2}}{v_i}. \tag{5}$$

T is a function of $n - 1$ variables, x_1 through x_{n-1}. Both x_0 and x_n are fixed, as are all the y_i's. To find the critical points for T, set each of the $n - 1$ partial derivatives of (5) equal to zero:

$$\frac{\partial T}{\partial x_i} = \frac{x_i - x_{i-1}}{v_i \sqrt{(x_i - x_{i-1})^2 + (y_i - y_{i-1})^2}}$$

$$- \frac{x_{i+1} - x_i}{v_{i+1} \sqrt{(x_{i+1} - x_i)^2 + (y_{i+1} - y_i)^2}} = 0,$$

so

$$\frac{x_i - x_{i-1}}{v_i \sqrt{(x_i - x_{i-1})^2 + (y_i - y_{i-1})^2}} = \frac{x_{i+1} - x_i}{v_{i+1} \sqrt{(x_{i+1} - x_i)^2 + (y_{i+1} - y_i)^2}}. \tag{6}$$

Since $\sin \alpha_i = (x_i - x_{i-1})/\sqrt{(x_i - x_{i-1})^2 + (y_i - y_{i-1})^2}$, (6) can be replaced by

$$\frac{\sin \alpha_i}{v_i} = \frac{\sin \alpha_{i+1}}{v_{i+1}},$$

where $1 \leq i < n$, which we rewrite as

$$\frac{\sin \alpha_i}{v_i} = c \tag{7}$$

where c is a constant, for all i. That is, T is minimized when (7) occurs. Therefore by (1), a shock wave traveling from A to B does so in a path of least time, which is Fermat's principle. Furthermore, (7) is true regardless of the number of layers between A and B. We state this as Snell's Law:

$$\boxed{\frac{\sin \alpha}{v} = c,} \tag{8}$$

Snell's Law

where for any point $P = (x, y)$ along a signal's path, v is a function giving the signal's speed at depth y, α is the angle of inclination of the tangent line to the path at P away from the vertical, and c is a constant, as in Figure 7a.

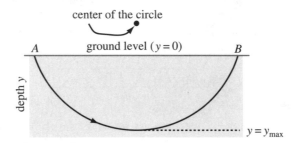

Figure 8. A path through a flat, affine earth

Let us transform (8) into a differential equation so that we can find the paths taken by waves through the earth. In Figure 7a, at $P = (x, y)$, construct the tangent line. So, $\cot \alpha = \frac{dy}{dx} = y'$. By (8) and with the help of Figure 7b, this becomes

$$c v = \sin \alpha = \frac{1}{\sqrt{1 + (y')^2}},$$

which, since $\frac{dx}{dy}$ is the reciprocal of $\frac{dy}{dx}$,

$$\boxed{\frac{1}{y'} = \frac{dx}{dy} = \pm \frac{c v}{\sqrt{1 - c^2 v^2}}.} \tag{9}$$

Snell's law as a differential equation

If the wave is penetrating into the earth, (i.e., moving in the positive direction), then y' should be non-negative, whereas if the wave is ascending towards the surface, (i.e., moving in the negative direction), then y' should be non-positive. The intuitive reason for the $\pm$ sign in (9) is that the paths that the signals take through the earth are reversible: if a signal travels along the curve Γ in going from A to B, then another signal could use the same curve Γ in going from B to A. These curves are two-way highways!

Example 1. To experiment with this version of Snell's law, let us solve (9) when velocity varies affinely with depth from the surface. That is, we launch a shock wave signal from the surface of a flat earth wherein the velocity of the signal at depth y is $v(y) = a + by$ where $y \geq 0$ and a and b are constants with $a > 0$ and where $v(y) > 0$. We take the initial point of the path as $A = (0, 0)$. Thus the positive alternative of (9) becomes

$$\frac{1}{y'} = \frac{dx}{dy} = \frac{c(a + by)}{\sqrt{1 - c^2(a + by)^2}}. \tag{10}$$

Because of the radical in (10), y' will be zero if y is $y_{\max} = (1 - ac)/(bc)$. Furthermore y can never exceed $y_{\max}$, because y' would no longer be real.

Solving (10) gives

$$x(y) = \frac{\sqrt{1 - a^2c^2} - \sqrt{1 - c^2(a + by)^2}}{bc}, \tag{11}$$

for $0 \leq y \leq y_{\max}$. When y attains the value $y_{\max}$, the wave can descend no further. Nevertheless it is still moving with speed $v(y_{\max})$ parallel to the x-axis. To find the rest of our wave's path we could take the negative alternative in (9) and use as the new initial condition $(2x_1(y_{max}), 0)$ where $x_1(y)$ is our original solution. (A computer algebra system will balk at the initial condition $(x_1(y_{max}), y_{max})$.) Alternatively, we could remember that the paths are two-way highways, and so realize that the second half of the path is obtained from the first half by a horizontal translation followed by a reflection in the y-axis. (We belabor this discussion, because we will encounter the same phenomenon when looking at the earth as comprised of concentric annuli rather than parallel layers.) However, a simpler way to find the second half of the path is to see that the paths are circles.

Rearranging (11) gives

$$\left(x - \frac{\sqrt{1 - a^2c^2}}{bc}\right)^2 + \left(y + \frac{a}{b}\right)^2 = \frac{1}{b^2c^2}, \tag{12}$$

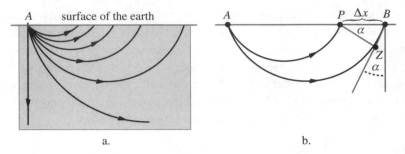

Figure 9. A wave front emanating from A

which we recognize as a circle of radius $1/(bc)$ and center $(\sqrt{1 - a^2c^2}/(bc),$ $-a/b)$, as shown in Figure 8.

To use this in practice, a, the speed at which shock signals travel through surface rock, can be determined by experiment. To find c we use a trick, as illustrated in Figure 9b. Imagine that points B and P host nearby seismograph posts. They each record an earthquake event at A and observe time difference Δt in their recorded times. From point P drop a perpendicular to the tangent line for the signal path AB corresponding to the tangency point B, so forming a right triangle $\triangle BZP$. Let α be the angle of inclination of BZ, so $\angle BPZ = \alpha$. Since B and P are very close, arcs AZ and AP have about the same length and shock signals traverse them in approximately equal time. If their velocity is a, then $a\,\Delta t/\Delta x \approx \sin\alpha = a\,c$, so

$$c \approx \frac{\Delta t}{\Delta x}.$$

With this value of c, (12) gives us a way to calculate b. If we know a, c, A, and B for a particular path, then the x-coordinate of the midpoint of segment AB must be $\sqrt{1 - a^2c^2}/(bc)$, which lets us find b. Once we know b, then for any signal path AB we can determine its associated c value by (12). When the initial distance between A and B is small, say less than 100 miles, this technique is used to help establish shock signal speed near the earth's surface, because an arc along 100 miles at earth's surface is fairly flat.

Through an earth of concentric annuli

Let us adapt Snell's Law to concentric layers or annuli. In Figure 10a we have a circle of radius r_1 within a circle of radius r_2, both having center O. Let A,

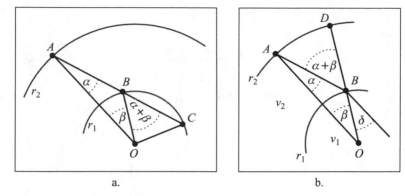

Figure 10. Adapting Snell's law to the circle

B, and C be collinear points on the circles as indicated. Let $\angle OAB = \alpha$ and $\angle AOB = \beta$. Therefore

$$\angle OBC = \alpha + \beta, \tag{13}$$

so $\angle BCO = \alpha + \beta$, since $\triangle OBC$ is isosceles. By the law of sines,

$$\frac{\sin \alpha}{r_1} = \frac{\sin(\alpha + \beta)}{r_2}, \tag{14}$$

so

$$r_2 \sin \alpha = r_1 \sin(\alpha + \beta). \tag{15}$$

Now modify Figure 10a into Figure 10b. The inner circle has radius r_1 and the outer circle has radius r_2. Suppose that signals travel with speed v_1 within the inner circle and with speed v_2 between the circles. Let points O, B, and D be collinear. Let the angles α, β, and δ be as indicated. By (13), $\angle ABD = \alpha + \beta$. By Snell's law (8)

$$\frac{r_1 \sin(\alpha + \beta)}{v_2} = \frac{r_1 \sin \delta}{v_1}, \tag{16}$$

which by (15) gives

$$\frac{r_2 \sin \alpha}{v_2} = \frac{r_1 \sin(\alpha + \beta)}{v_2}. \tag{17}$$

By (16) and (17), we have

$$\frac{r_2 \sin \alpha}{v_2} = \frac{r_1 \sin \delta}{v_1}. \tag{18}$$

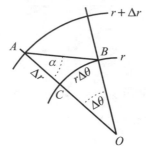

Figure 11. Translating Benndorf's law into a differential equation

By the same process used in obtaining (8) from (7), we obtain Benndorf's law (also known as Bouguer's formula) from (18), the adaptation of Snell's law to circular boundaries:

$$\boxed{\frac{r \sin \alpha}{v} = c.} \tag{19}$$

Benndorf's law

That is, for any given signal there is a constant c for which (19) is valid at every point A along the signal's path, where v is the speed of the signal at radial distance r from O and α is the angle of inclination of the signal's path away from ray OA.

To transform (19) into a differential equation, in Figure 11 assume that in the annular region r to $r + \Delta r$, the speed v of a signal is constant, where Δr is a small change in the radial distance from the origin O. When Δr is small, then the path $ABCA$ consisting of line segments and arcs of circles is approximately a right triangle. Since $\sin \alpha \approx r\Delta\theta/\sqrt{(\Delta r)^2 + (r\Delta\theta)^2}$, Benndorf's relation becomes

$$\frac{r^2 \Delta\theta}{v \, \Delta r \sqrt{1 + (r \frac{\Delta\theta}{\Delta r})^2}} = c. \tag{20}$$

As Δr approaches 0 in (20), we get

$$\theta' r^2 = cv\sqrt{1 + (r\theta')^2}, \tag{21}$$

where $\theta' = \frac{d\theta}{dr}$. Rearranging gives

$$\boxed{\theta' = \pm \frac{cv}{r\sqrt{r^2 - c^2v^2}}.} \tag{22}$$

Benndorf's law as a differential equation

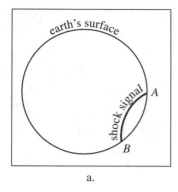
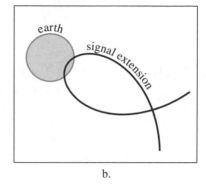

a.   b.

Figure 12. A signal through the earth

Example 2.   Imagine an earth of radius $R = 2$ where the density function $v(r)$ decreases as the radial distance r from the earth's center increases, a reasonable assumption since we expect the earth to be more dense near its center than at its surface. Let us take $v(r) = 3 - r$ and $c = 1$. Take the starting point of our signal as $A = (2, 0)$ in polar coordinates. With these values, (22) becomes

$$\theta' = \frac{3 - r}{r\sqrt{6r - 9}}, \tag{23}$$

taking the positive square root. As r decreases, the signal penetrates into the earth, and so $\frac{dr}{dt}$ is negative. Since $\frac{d\theta}{dr}$ must be positive by (23), then $\frac{d\theta}{dt}$ is negative, which means that the signal will venture into the fourth quadrant. Because of the radical in (23) the nearest approach to the earth's center is $r_{\min} = 1.5$. The solution for (23), $r_{\min} \le r \le 2$, is

$$\theta(r) = 2\tan^{-1}\left(\frac{\sqrt{6r - 9}}{3}\right) - \frac{\sqrt{6r - 9} - \sqrt{3} + \pi}{3}.$$

Once the signal dips to this minimum allowed r value, we use the negative square root in (22) to generate the second half of the signal's path, just as we did for the example of a signal through a flat earth. At first glance, the generated signal in Figure 12a looks like the arc of a circle. However if we allow the solution curves to extend backwards above the surface of the earth (still faithfully following the speed function v, even though v becomes negative beyond $r = 3$), Figure 12b shows that the curve ultimately crosses itself, and hence by analytic continuation the original path cannot be the arc of any circle.

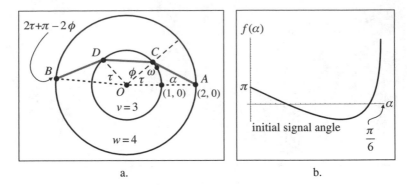

Figure 13. Signals through the mantle and core

Limits of the Shadow

When earthquakes occur, shock signals are detected by seismographs around the world except by those located within the shadow regions. For a simple model of the earth, let's determine the extent of this shadow.

Example 3. Imagine that the earth has radius 2 and that it has an inner liquid core of radius 1. Position the origin O at the earth's center. Suppose an earthquake has occured at $A = (2, 0)$ in polar coordinates, as displayed in Figure 13a. Let α be the angle between ray OA and a signal emanating from A, where $-\frac{\pi}{2} \leq \alpha \leq \frac{\pi}{2}$. (For any other α value, the signal is lost in the atmosphere.) For simplicity, assume that the signal speed within the core is 3, and within the mantle (the region between the two circles) it is 4. Figure 14 is an adaptation from [**51**, p. 154] showing signal speed within the earth; contrary to what we might expect, the signal speed fails to decrease monotonically with distance from the earth's center; in fact, within earth's outer liquid core it is much less than in the surrounding mantle. We mirror this phenomenon in our simple model.

For any signal from A with $\pi/6 < \alpha < \pi/2$, the signal's path will be a chord lying entirely within the mantle. If $0 \leq \alpha \leq \pi/6$, the signal will penetrate the core. For any such α, we wish to determine where it will emerge from the earth. The signal strikes the core at C. Let $\omega = \angle ACO, \tau = \angle AOC$, and $\phi = \angle DCO$. By the law of sines, $\sin \alpha / 1 = \sin \omega / 2$. Since ω is obtuse,

$$\omega = \pi - \sin^{-1}(2 \sin \alpha),$$

so

$$\tau = \sin^{-1}(2 \sin \alpha) - \alpha. \tag{24}$$

By Snell's law, $\sin(\pi - \omega)/4 = \sin\phi/3$, which gives

$$\phi = \sin^{-1}\left(\frac{3\sin\alpha}{2}\right). \tag{25}$$

Thus by (18) and (25) the signal reaches the earth's surface at central angle

$$f(\alpha) = \pi + 2(\tau - \phi)$$
$$= \pi + 2\left(\sin^{-1}(2\sin\alpha) - \alpha - \sin^{-1}\left(\frac{3\sin\alpha}{2}\right)\right), \tag{26}$$

where $\alpha \le \pi/6$. For $\pi/6 < \alpha \le \pi/2$, the signal surfaces at central angle $f(\alpha) = \pi - 2\alpha$. Figure 13b gives a graph of surface angle $f(\alpha)$ versus initial signal angle for $0 \le \alpha \le \pi/6$. In Exercise 6, the reader is invited to find the angle for which $f(\alpha)$ is a minimum.

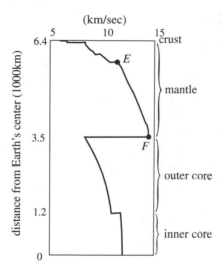

Figure 14. Signal speed in the earth

Figure 15 shows a family of shock signals emanating from A. No signal reaches the earth's surface between points B and C and between points E and F. Point B arises from $\alpha = \frac{\pi}{6}$, a signal path tangent to the inner circle, whereas point C arises from the signal at $\alpha = -\frac{\pi}{6}$, which enters into the fourth quadrant rather than the first quadrant. For this signal, $\phi = \sin^{-1}\frac{3}{4}$. By (26), the shadow extends from 120° to about 157°.

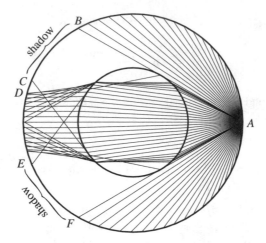

Figure 15. Shadow cast by the core

Finding the depth of the core

This same shadow phenomenon occurs when the speed function for signals within the mantle is an affine function, except that now the signal paths within the mantle look more like circular arcs than lines. There will be a signal from the earthquake site at A whose path is tangent to the core, as in Figure 16. Just as in Figure 15, the shadow starts at B. We will use this path to determine the depth of the core. From observation we can determine Θ, the central angle from A to B, and we can determine the speed at which signals travel near the surface of the earth.

To find c of (19) we adapt the trick of Figure 9b, also found in [**84**, p. 96]. In Figure 17, imagine an earthquake occuring at A and being recorded at B, the start of the shadow. (In practice, we can set off an explosive at A.) Suppose

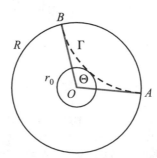

Figure 16. Tangent signal path to the core

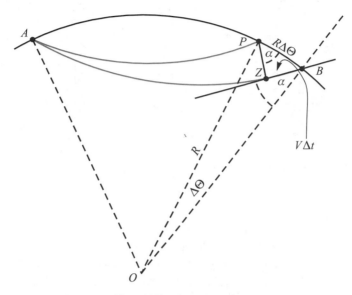

Figure 17. A trick to find c

we also record the event from a nearby post P. Drop a perpendicular from P to the tangent line for the signal path AB corresponding to tangency point B. Angles OQP and PQZ are complementary, hence $\angle OQP = \alpha = \angle QPZ$. Let V be the signal speed near R, the radial distance of the earth's surface from O. The paths AP and AZ are approximately the same length and so are traversed by signals in about equal times. Let Δt be the time difference between the signal being heard at post P and B. Then $\sin \alpha = V \Delta t / (R \Delta\Theta)$, and by Benndorf's law, $\sin \alpha = c V / R$. Thus

$$c \approx \frac{\Delta t}{\Delta \Theta}. \tag{27}$$

Example 4. To estimate the depth of the core, assume that the radius of the earth is $R = 6400$ km, that the signal speed within the mantle is $v(r) = a + br$, and that the signal speed within the core, as indicated in Figure 14, is less than in the lower mantle, so creating a shadow. Let Γ be the tangent signal path from A to B as in Figure 16. In 1914, B. Gutenberg discovered that the central angle to the endpoints of Γ was 105°. Imagine we have a listening post P which is 10 km from B, as indicated in Figure 17. Furthermore, suppose that the time difference between hearing the earthquake from A is $\Delta t = 0.5$ seconds at posts B and P. Thus by (27), $c \approx 320$. In order

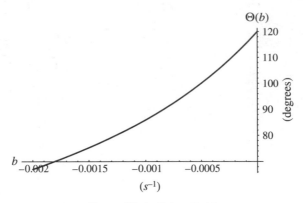

Figure 18. A trick to find b

to get a good estimate for the location of the core/mantle boundary, we need a reasonable value for the surface speed of shock signals. From Figure 14, we probably should pick the value 6. However, if we do so, such a low speed value at the surface versus the steep speed change in the lower mantle will cause Γ to plunge deeper than we want. Instead, let us take the surface signal speed to be the intersection of the line determined by segment EF and the earth's surface in Figure 14. That is, choose the surface signal speed as about 10 km/sec. By Benndorf's law, $6400 \sin\alpha/10 = 320$ means that the initial angle of inclination of Γ away from OA is $30°$. From (22), the radial depth of the core-mantle boundary is $r_0 = ac/(1 - bc)$, which occurs when the radical in (22) is 0. Since $v(R) = 10$, $a = 10 - bR$. From (22),

$$\Theta(b) = 2 \int_{r_0}^{R} \frac{c\,v}{r\sqrt{r^2 - c^2 v^2}}\, dr, \tag{28}$$

a function of b, since r_0 and a are functions of b. Figure 18 shows a graph of (28). For $b \approx -.00036$, $\Theta \approx 105°$, which means that $a \approx 12.30$ and $r_0 \approx 3530$ km, which is very close to the actual value of 3511 km.

If you object that the above example has been rigged by using artificially good guesses for Δt and surface signal speed $a + bR$, you are right, of course. Nevertheless, it demonstrates the power of Snell's law.

Example 5.   For a more realistic execution of the method, from Figure 14, we see that the signal speed in the mantle near the core-mantle boundary is about 14 km/sec. For the shock signal Γ in Figure 16, when $r = 3511$ km (the core-mantle boundary), $\alpha = \pi/2$. Thus, by Benndorf's law, the approximate value of c for Γ should be about 252, being the value of $3511 \sin(\pi/2)/14$.

Furthermore, take the surface signal speed as 6. Repeating the computations as indicated above gives 2647 km, not quite 900 km too low.

With a little more work we could improve the method, using a two-fold piecewise affine function to approximate the speed within the mantle and using the time lapse for the signal to travel Γ, to achieve a close approximation. With a lot more analysis, the signal speed profile of Figure 14 has been worked out in detail by geophysicists over the years. Their analysis is a convoluted yet monumental achievement. This chapter gives a start to the reader interested in the rest of the story.

Exercises

1. Assume a flat earth, that $v(y) = 1 + y$ where y is depth in hundreds of miles, and that $v(y)$ is hundreds of miles per minute. At a listening post 200 miles from the blast, Helen's seismograph first registers shock waves from the blast at 10:00 am. Let Γ be the path taken by the registered signal. How deep does Γ go? When did the tremor occur?

2. Show that the time lapse T along a signal path from r_1 to r_2 in a round earth is given by

$$T = \int_{r_1}^{r_2} \frac{r}{v\sqrt{r^2 - c^2 v^2}} \, dr. \qquad (29)$$

3. Use (2) to determine the time lapse for a signal to travel Γ in example 4 where $c = 320$, $a = 12.30$, and $b = -.00036$.

4. Construct a piecewise affine function from Figure 14 giving the speed of the shock signals r km from the center. Use this function to calculate the time for a signal to traverse a path starting out like the one in Example 4. That is, $c = 320$ and $v(R) = 10$. Compare your answer with Exercise 3.

5. Suppose that the angle from A to B was $90°$ in Example 4. How deep is the core-mantle boundary?

6. In (26) find the positive angle α_1 for which f is a minimum. Let Γ_1 be the signal path associated with α_1. Let $\alpha_2 = -\pi/6$, and let Γ_2 be the signal path associated with α_2. Find the common point P within the mantle lying on both Γ_1 and Γ_2. Determine the time along both Γ_1 and Γ_2 from A to P. Conclude that Snell's law when applied to nonparallel mediums does not always result in a minimum time path between two points.

7. In Example 3, change the speed within the mantle to $v(r) = 5 - r$. Generate and display a family of shock paths from an earthquake at A, obtaining a figure analogous to Figure 15.

8. Imagine that a shock signal proceeds from the polar point $A = (2, 0)$ in a ball of radius $R = 2$ with initial angle $\alpha = \pi/6$ and that $v(r) = r/4$. Plot the signal's path. How deep does the signal go?

9. The Little Prince's planet as discussed in Chapter XI, Asteroid B-612, has volcanic eruptions, [24]. His asteroid has diameter 35 feet. A volcano has erupted, shooting forth fireballs at frequent intervals, so causing shock waves through his planet. He finds that there is a shadow region between $100°$ and $120°$ along a great circle through the volcano. Assume that the surface signal speed is 6 ft/sec, and that the initial angle of the tangent signal is $\alpha = 20°$. Assume that the mantle signal speed is affine. How deep is the core-mantle boundary? How fast is the signal speed in the lower mantle? When the little prince puts his ear to the ground, he can hear the fireballs erupting from the antipodes five seconds after the event. Assume the core is homogeneous. Can any conclusions be made about the core signal speed?

10. Here is an integral which arises from the context of the material in this chapter. It is a tidied up version of (2) where $v = \sqrt{4 - x^2}$. Can you crack it?

$$\int_1^2 \frac{2x}{\sqrt{4 - x^2}\sqrt{x^2 - 1}} \, dx$$

preamble x

Make me, O Lord, thy spinning wheel complete.

Edward Taylor (1644–1729), *Housewifery*

Ten applications for graduate school lay on my desk. I was in my senior year at a small liberal arts school in the Midwest. Eight other graduating math majors were in my class, but none of them were going on for further studies in pure mathematics just yet. Was I crazy? Scanning through the forms and hoping for some inspiration as to which one to fill out first, I noticed that all but one required a processing fee of $50. My bank account was empty, so the decision was easy. I completed the free form and sent it off—to Wyoming—and let the others gather dust.

The algorithms we use for life are often nonlogical. I had just made a momentous decision in a Brownian motion kind of way in which objects bounce off one another in random fashion. My advisor had earned his degree from Wyoming. Beyond that I imagined that one graduate school was as good as another. I was just beginning to learn.

However, even if one uses mostly logical algorithms in life, life is serendipitous. Who knows what may happen? Let me illustrate.

A year behind me at my alma mater was another math major for whom I was a role model. He wound up in graduate school in Colorado, just over the border from Wyoming. From time to time we visited each other, comparing notes, comparing programs, and hiking the territory. Between us was a mountain ridge. One fine Saturday morning in the early summer I drove off to visit my friend. Weather in the high country can be unpredictable. Snow began to fall as the elevation increased. Further on, inches of snow lay on the road over a layer of ice. This can't be happening, it's the July Fourth weekend! About the time I was admitting defeat and ready to turn around, my tires spun out beneath me. My Pinto began a slow turn about its center of mass while skating up the highway. Ahead of me, in an oncoming red sedan, I saw the face of its driver. His eyes grew wide. How could he avoid me? Logic dictated disaster. When we passed each other our vehicles were parallel, both pointing north. I continued a slow pirouette, completing a loop. I could never duplicate the feat intentionally. Nevertheless I had just traced the petal of an epitrochoid of sorts, as I seem to keep doing as the years go by, spinning in semi-confusion while trying to figure out what's going on.

CHAPTER X

The Trochoid Family

Some of the most beautiful curves studied in calculus are members of the trochoid family. In the mathematical zoo of the typical calculus text, they are usually presented as the curves that are traced by a tack embedded somewhere in a wheel that is rolling around a circle. In Figure 1a, the dotted railroad wheel has a tack in its rim which overlaps the dashed circle road; the solid curve is the trochoid, the path traced out by the tack. In Figure 1b, the wheel is rotating on the inside of the circle. In this chapter, we give a little history, showing how the trochoids are related to falling and spinning motion. We then present them using both the traditional wheel model and a more unusual model where the trochoid is the curve traced out by a tack in a bungee cord whose ends are held by two walkers proceeding around a circular track. If we imagine the cord as being made of glowing material, then a time-lapse taken from above, stroboscopic photo of the action leads us to the notion of

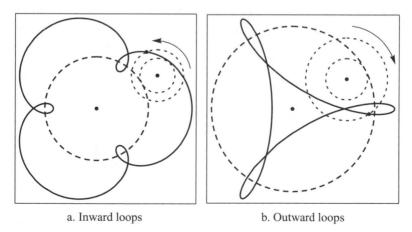

a. Inward loops   b. Outward loops

Figure 1. The trochoids

the envelope to a curve and then on to fractional derivatives and harmonic envelopes.

A little history

As the ancients of the northern hemisphere tracked the planets across the fixed background of stars, they noticed that their counterclockwise motion about the earth was interrupted by periods of clockwise motion.

For example, consider Figure 2, where E, J, and S represent the positions of the earth, Jupiter, and a fixed star, respectively, at times $t = 0$, $t = 0.5$ years, and $t = 1$ year.

We assume that S remains fixed and that E and J move counterclockwise in circular orbits of radii 1 and 5 and periods 1 and 12. In Figure 2a, E is ahead of J with respect to S, and six months later, in Figure 2b, E is behind J. In another six months, in Figure 2c, E is again ahead of J. Earth's position is given by

$$E(t) = (\cos(2\pi(t + .25)), \sin(2\pi(t + .25))) \tag{1}$$

and Jupiter's position is given by

$$J(t) = 5\left(\cos\left(\frac{2\pi}{12}(t - .8)\right), \sin\left(\frac{2\pi}{12}(t - .8)\right)\right), \tag{2}$$

where the phase shifts of 0.25 years and -0.8 years have been chosen for convenience.

Until the time of Copernicus, the accepted model of the solar system was the geocentric model of Ptolemy. See Figure 2 in Chapter I for a medieval rendition of Ptolemy's model. In that model, let us suppose that the earth is fixed, and then plot the position of Jupiter with respect to earth. We use

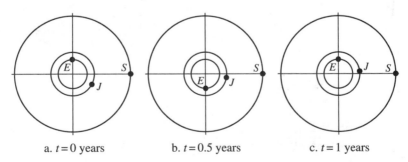

a. $t=0$ years   b. $t=0.5$ years   c. $t=1$ years

Figure 2. Earth-Jupiter-Fixed Star snapshots

(1) and (2) to find the distance $r(t)$ of Jupiter from the earth at time t and the angle $\theta(t)$ by which earth is ahead of Jupiter, where $-\pi \leq \theta(t) \leq \pi$ for all t. The distance is $r(t) = ||J(t) - E(t)||$. To find $\theta(t)$, use (3) of the appendix for finding the angle between the unit vector $\mathbf{i}$ and $J(t) - E(t)$, with the proviso that $\theta(t)$ is positive if and only if the second component of $E(t) - J(t)$ is positive. That is,

$$r(t) = ||J(t) - E(t)|| \quad \text{and}$$

$$\theta(t) = \cos^{-1}\left(\frac{J_1(t) - E_1(t)}{r(t)}\right) \text{Sign}(E_2(t) - J_2(t)). \tag{3}$$

The polar plot of $(r(t), \theta(t))$ in Figure 3a for the first year is how an observer on earth sees Jupiter's location in the sky with respect to the fixed star S. (We include the portion of the path when the sun is between the two planets even though an observer could not see it.) What the ancients noticed is that sometime between $t = 0.5$ years and $t = 1$ year, Jupiter appears to be traveling clockwise, or to be going backwards. Anyone walking in a forest experiences a similar phenomenon, for as a hiker walks along, the nearby tree trunks seem to be sliding backwards against the backdrop of the tree trunks far away.

Running this polar model (3) over a period of twelve years gives the epitrochoidal-like curve of Figure 3b. The ancients used relative brightness in determining relative distance of a planet to earth over time, a notion that supported the idea that the planets moved in their realms following paths like Figure 3. Unfortunately, earth's and Jupiter's orbits about the sun are only approximately circles, so it was impossible for ancient astronomers to model their perception of reality. In fact, when Copernicus proposed a heliocentric

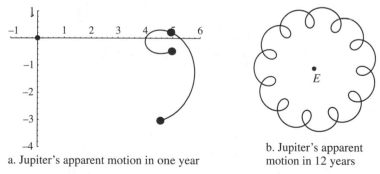

a. Jupiter's apparent motion in one year

b. Jupiter's apparent motion in 12 years

Figure 3. Jupiter's apparent epicycloidal path about earth

system, he borrowed the notion of epicycles from the Ptolemaic system, and used 34 circles to characterize the motion of the six planets and the moon [**1**, p. 79].

Despite the apparent complication of planetary motion away from that of circular motion, the ancients assured themselves that this motion, in Plutarch's words, was

> not attributable to irregularity or confusion; but in them astronomers demonstrate a marvellous order and progression, making [the planets] revolve with circles that unroll about other circles, [so that the planets] retrogress smoothly and regularly with ever constant velocity [**63**, pp. 167–9].

However, the arclength along the curve of Figure 3b is

$$L(t) = \int_0^t \sqrt{(r'(\tau)^2 + (r(\tau)\theta'(\tau))^2}\, d\tau \text{ AU},$$

where an AU is an astronomical unit, the distance of the earth from the sun. Thus the speed at which the apparent path of Jupiter is traced is $L'(t)$, whose graph is given in Figure 4. That is, the speed of Jupiter along its apparent path in the heavens is nowhere near constant!

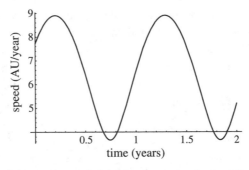

Figure 4. Jupiter's speed along its apparent path about the earth

The wheel model for the trochoids

Using the wheel model, let us derive parametric equations for a trochoid in terms of a variable t, that we refer to as time. Take a tack of glowing material and embed it in a wheel that rolls in the xy-plane around a fixed circle whose center is the origin O and whose radius R is a positive real number. The

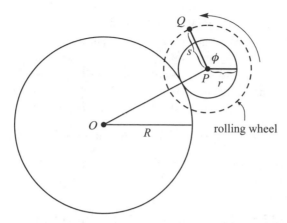

Figure 5. The wheel setup

distance r from the center of the wheel to the road we suppose is non-zero. To allow for some normalized standard equations, let us say that the initial position of the glowing tack is on the x-axis at a point other than the origin. Let P and Q be the positions of the center of the wheel and of the tack, respectively. The center of the wheel is initially at the point $P_0 = (R + r, 0)$ and the tack is s units from the wheel's center so that the tack is initially at $Q_0 = (R + r + s, 0)$, as in Figure 5. Since s may exceed $|r|$, the wheel may at times look like a railroad wheel whose flange extends beyond the upper surface of the track. We interpret the wheel as rolling around the outside of the circle of radius R if $r > 0$, and as rolling around the inside if $r < 0$; furthermore, we interpret the tack as initially being on the right-hand side of the wheel if $s > 0$, and on the left-hand side if $s < 0$.

The path of the glowing tack traced out by Q is called a trochoid when its motion is periodic in time, which happens when R/r is a non-zero rational number m. Let θ be the angle between the positive x-axis and a ray from O through P. Let ϕ be the angle between the direction $Q_0 - P_0$ and the direction $Q - P$. So Q is given by the vector sum of the vector from the origin to the wheel's center and the vector from this center to the tack; that is, the sum $(R + r)(\cos\theta, \sin\theta) + s(\cos\phi, \sin\phi)$, where initially $\theta = 0 = \phi$. Figure 5 shows the case when $R > 0$ and $s > r > 0$. By the arclength formula, $R\theta = r\phi$. Hence Q can be written as $(R+r)(\cos\theta, \sin\theta) + s(\cos m\theta, \sin m\theta)$. Next, choose any non-zero rational number p and let $q = mp$. Using the reparametrization $pt = \theta$ gives Q as $r(1 + q/p)(\cos pt, \sin pt) + s(\cos qt, \sin qt)$. If $q/p = -1$, then Q's path is a circle of radius s. Otherwise, since any non-zero scalar multiple of this curve is simply a rescaling (along with a reflection in the case of a negative scalar), then any curve of the form $\alpha(\cos pt, \sin pt) + \beta(\cos qt, \sin qt)$ is a trochoid, where p and q are arbitrary rational numbers and α and β are real numbers with $\alpha + \beta \neq 0$. We standardize these equations by dividing by $\alpha + \beta$ so that the tack is initially at $(1, 0)$. Now let $\lambda = \alpha/(\alpha + \beta)$ so that $1 - \lambda = \beta/(\alpha + \beta)$. Hence the standard classic equations for a trochoid are

$$(x, y) = (\lambda \cos pt + (1 - \lambda) \cos qt, \lambda \sin pt + (1 - \lambda) \sin qt), \quad (4)$$

where p and q, $p \neq -q$, are rational numbers and λ is any real number. The path is symmetric with respect to the x-axis and $(x(0), y(0)) = (1, 0)$.

The bungee cord, an alternate trochoid model

Suppose that two people A and B walk along holding a taut bungee cord between them. We assume that this bungee cord stretches uniformly and can collapse to a point. Embed a glowing tack in the bungee cord; a time-lapse photo will trace out a curve as A and B walk along. For example, in Figure 6, A and B walk counterclockwise around the unit circle, with A walking five times as fast as B while the tack stays in the middle of the bungee cord. The line segment in the figure represents the bungee cord at the particular time $t = 0.5$, while the path within the unit circle is the path of the tack. In general, if A makes p circuits of the unit circle while B makes q circuits, where p and q are rational numbers, then at time t, A's position is $(\cos pt, \sin pt)$ and B's position is $(\cos qt, \sin qt)$. Now embed the tack somewhere in the bungee cord between A and B, and let λ be the ratio of the length between the tack and B and of the length between A and B, (where $\lambda = 1$ if A and

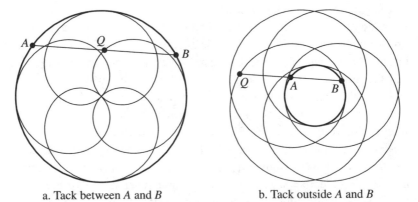

a. Tack between A and B b. Tack outside A and B

Figure 6. Following the tack

B have the same position). Then the tack's position at time t is given by $(x, y) = (\lambda \cos pt + (1 - \lambda) \cos qt, \lambda \sin pt + (1 - \lambda) \sin qt)$, where λ is a real number between 0 and 1.

For a more general case, we envision the two walkers as holding onto a point of an infinite bungee line (rather than just a cord) that stretches uniformly as the distance between the two varies. In this way, the tack can be positioned anywhere along the bungee line, not just between the two walkers. For example, Figure 6b gives the trochoid for $p = 5, q = 1, \lambda = 2$, where the line represents the bungee cord at $t = 0.5$. Here the path of the tack is a trochoid outside the unit circle. The standard bungee cord trochoid parametrizations are

$$
\begin{aligned}
T(p, q, \lambda) &= (x, y) \\
&= \big(\lambda \cos pt + (1 - \lambda) \cos qt, \lambda \sin pt + (1 - \lambda) \sin qt\big),
\end{aligned}
\tag{5}
$$

where p and q are any rational numbers with at least one of p and q being non-zero and λ is any real number. If we allow $q = -p$ and $\lambda = 1/2$, then the path generated by (5) is a line segment. The difference between (1) and (5) is that (5) allows any nontrivial ellipse to be considered as a trochoid.

A more stunning way to perceive the trochoids is to imagine that A and B travel about separate unit circle tracks, one a unit distance above the other; if the cords themselves are made of glowing material, a time-lapse photo of the cord in motion yields a surface in $\mathcal{R}^3$, and each horizontal cross section

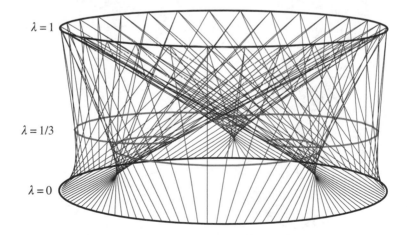

Figure 7. Layers of trochoids, $p = 4, q = 1$

is a trochoid. A parametrization of this surface is

$$G(t, \lambda) = (\lambda \cos pt + (1 - \lambda) \cos qt, \lambda \sin pt + (1 - \lambda) \sin qt, \lambda), \quad (6)$$

where $0 \le t \le 2\pi$ and λ is a real number. $G(t, 1)$ and $G(t, 0)$ are the two unit circle tracks on which A and B are walking. Figure 7 gives a stroboscopic time-lapse photo of this surface in which a few of the bungee cords are illuminated and the solid gray curve is the cross section for $\lambda = 1/3$.

Envelopes

Overhead shots of Figure 7 yield striking views, as in Figure 8. That is, when squashed back into the plane, the bungee cords at any time t appear to be tangent to a curve we call C, as in Figure 8d. There are two natural ways to view C. From a global perspective, C is where the surface of Figure 8 would fold upon itself when squashed down into the plane. From a local perspective, (x, y) is a point on C if (x, y) is on the bungee cord at some time t and (x, y) always stays on the same side of the curve C near time t.

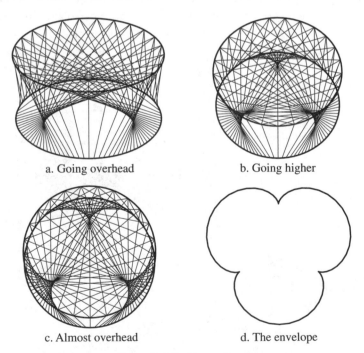

a. Going overhead   b. Going higher

c. Almost overhead   d. The envelope

Figure 8. Moving overhead

Let us consider the global interpretation first. If we squash or project the surface of Figure 8 onto the xy-plane so that each $G(t, \lambda) = (x, y, \lambda)$ is projected to $(x, y, 0)$, then the surface folds upon itself creating creases or curves, much like a heap of linen sheets pressed flat onto the floor. If (x, y) is on a fold then there should be a normal vector to the surface at (x, y, t) that is parallel to the xy-plane; that is, the third component of the normal vector should be 0. A normal to G can be found by calculating $G_t \times G_\lambda$, where G_t and G_λ are the gradients of G with respect to t and λ. That is,

$$G_t = \left(-\lambda p \sin pt - (1-\lambda)q \sin qt, \lambda p \cos pt + (1-\lambda)q \cos qt, 0 \right)$$

and

$$G_\lambda = (\cos pt - \cos qt, \sin pt - \sin qt, 1).$$

The third component of $G_t \times G_\lambda$ is

$$-\lambda p(\sin^2 pt + \cos^2 pt) + (1-\lambda)q(\sin^2 qt + \cos^2 qt)$$
$$+ (\lambda p - (1-\lambda)q)(\cos pt \cos qt + \sin pt \sin qt).$$

Setting that equal to 0, using some trigonometric identities, and solving for λ gives

$$\lambda = \frac{q}{p+q}.$$

Substituting into (5) gives the parametric equation for C as

$$\left(x(t), y(t) \right) = \left(\frac{p\cos(qt) + q\cos(pt)}{p+q}, \frac{p\sin(qt) + q\sin(pt)}{p+q} \right). \quad (7)$$

This is a very special case of (4). When p and q are both positive then (7) is a parametrization of the epicycloid (i.e., when $s = r$), and when p and q are of opposite signs then (7) is a parametrization of the hypocycloid (i.e., when $r < 0$ and $s = |r|$). Figure 9 shows the case when $p = 5, q = -1$.

The classic way to achieve this same result is to use a representation involving the points (x, y) on the bungee cord lines, rather than the convex combination representation of (4) as used above. To obtain this representation, we see that a point (x, y) is on the bungee line at time t if and only if

$$\frac{y - \sin pt}{x - \cos pt} = \frac{\sin qt - \sin pt}{\cos qt - \cos pt}.$$

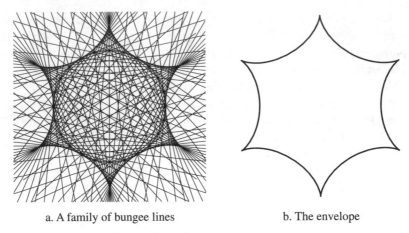

a. A family of bungee lines   b. The envelope

Figure 9. A hypocycloid

After rearranging terms and using a trigonometric addition identity, we see
that (x, y) is on the bungee cord at time t if and only if $F(x, y, t) = 0$ where

$$F(x, y, t) = x(\sin pt - \sin qt) + y(\cos qt - \cos pt) + \sin(qt - pt). \quad (8)$$

Provided t is not a time at which A and B occupy the same place,
$F(x, y, t) = c$ is a line parallel to the line $F(x, y, t) = 0$ in the xy-
plane. For such times t, we say that a point (x, y) is above the bungee
line if $F(x, y, t) > 0$, below the line if $F(x, y, t) < 0$ and on the line if
$F(x, y, t) = 0$.

The set of points (x, y, t) for which $F = 0$ is a surface in $\mathcal{R}^3$, typified by
Figure 10. When the surface $F = 0$ for $1 < t < 3$ of this figure is projected
onto the xy-plane, a crease or fold will be created. As before, we want to
determine the fold set of this surface when it is projected onto the xy-plane. A
point (x, y) is on the fold set whenever there is a tangent plane to the surface
at (x, y, t) for which the t-axis is parallel to the tangent plane. This means
that $F_t(x, y, t) = 0$, where F_t is the partial derivative of F with respect to t.
Hence the fold curve C, classically called the envelope of the family of lines
$F(x, y, t) = 0$ indexed by t, is obtained by solving the system

$$F(x, y, t) = 0 \quad \text{and} \quad F_t(x, y, t) = 0. \quad (9)$$

After straightforward simplification, the parametric solution is that in (7). See
[**73**] for details; example 3 of [**11**] gives the same result using an altogether

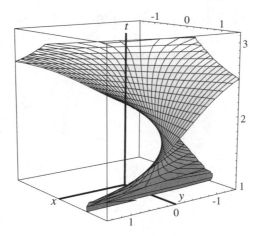

Figure 10. The surface $F = 0$ for $p = 2, q = 1, 1 < t < 3$

different approach. See [**15**, p. 76] for a more formal presentation of this definition of the envelope.

Now let us consider the local perspective: to find those points (x, y, t) for which (x, y) is both on the bungee cord at time t and for which the bungee cord always stays on the same side of (x, y) near time t, we need to identify those points (x, y, t) for which both $F(x, y, t) = 0$ and for which $F(x, y, s)$ is either solely nonnegative or solely nonpositive for s near t; that is, t is both a root of $F(x, y, s)$ and a local extreme of $F(x, y, s)$ with respect to s. Such points will be contained in the solution set to (9). Both the global and local perspectives lead us to solving the same system of equations!

Which trochoids are bungee envelopes?

Since the epicycloid and hypocycloid appear nicely as envelopes of bungee cords, it is natural to ask if the general trochoid is also the envelope of a similar family of lines. With this in mind we pose the following problem where f and g are functions from $\mathcal{R}^2$ into $\mathcal{R}$.

A Bungee Cord Characterization: Given a curve Γ in the x-y plane, find a function pair (f, g) such that the envelope of the family of lines $\{\mathcal{L}_t \mid t \in \mathcal{R}\}$ is Γ, where $\mathcal{L}_t$ is the line through $(\cos f(t), \sin f(t))$ and $(\cos g(t), \sin g(t))$.

For example, for epicycloids and hypocycloids, the function pairs consist of the linear functions (pt, qt) where p and q are different nonzero rational numbers with $p \neq -q$. In general, if A and B proceed at varying rates about the unit circle, then we have the following result.

Theorem 1 (A Parametrized Envelope) *The envelope formed by the family of lines whose endpoints are* $(\cos f(t), \sin f(t))$ *and* $(\cos g(t), \sin g(t))$ *where f and g are different differentiable functions is the parametric set of equations:*

$$
\big(x(t), y(t)\big) =
$$
$$
\left(\frac{f'(t)\cos(g(t)) + g'(t)\cos(f(t))}{f'(t) + g'(t)}, \ \frac{f'(t)\sin(gt) + g'(t)\sin(f(t))}{f'(t) + g'(t)} \right). \tag{10}
$$

To derive these equations note that a point (x, y) on any of these lines must satisfy

$$
y - \sin f(t) = \frac{\sin g(t) - \sin f(t)}{\cos g(t) - \cos f(t)}(x - \cos f(t))
$$

for some t. Rearranging terms, we can represent the set of points (x, y, t) satisfying such equations as the zero level set of the function

$$
F(x, y, t) = x\big(\sin(f(t)) - \sin(g(t))\big)
$$
$$
+ y\big(\cos(g(t)) - \cos(f(t))\big) + \sin\big(g(t) - f(t)\big).
$$

Hence

$$
F_t = x\big(f'(t)\cos(f(t)) - g'(t)\cos(g(t))\big)
$$
$$
+ y\big(f'(t)\sin(f(t)) - g'(t)\sin(g(t))\big)
$$
$$
+ \big(g'(t) - f'(t)\big)\cos\big(g(t) - f(t)\big).
$$

After straightforward simplification, similar to the steps used in solving (9), the solution set to the equations $\{F = 0, F_t = 0\}$ can be written in the desired form.

In terms of the bungee cord characterization problem, we would like to find that function pair (f, g) that yields a given trochoid curve Γ as the envelope. Since these trochoid curves are so strongly related to the epicycloid and hypocycloid, one might guess that finding the bungee cord characterizations for a typical trochoid is easy. Let Γ be the graph of the cardioid $r = 1/2 + \cos\theta$ shifted to the left by $1/2$. A poor first guess bungee cord characterization of Γ appears in Figure 11a where $f(t) = 2t + 2\sin 2t$ and $g(t) = t$; Γ is the solid curve and the envelope is the dashed curve. One near miss for a bungee cord characterization of Γ is the function pair $(f, g) = (2t + \frac{\pi}{2}\sin\frac{t}{2}, 2t - \frac{\pi}{2}\sin\frac{t}{2})$, but as one can see from Figure 11b, the envelope (the dotted curve) is not quite Γ (the solid curve). As far as we

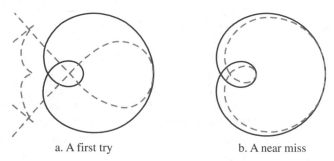

a. A first try　　　　　　　　b. A near miss

Figure 11. Bogus trochoids

know, finding the bungee cord characterization for the trochoids remains an open problem.

Harmonic envelopes

Although we are unable to produce the general trochoid as an envelope of a family of bungee cord lines held by two walkers moving around the unit circle, we can generate trochoids using the higher order partial derivatives of F. Since $F = 0$ can be perceived as a surface in $\mathcal{R}^3$, so can $\frac{\partial^u F}{\partial t^u} = 0$. In fact, such surfaces will look very much like Figure 10. The intersection of any two will be a curve in $\mathcal{R}^3$. With this idea in mind, we define $\mathcal{F}_u$, called the uth harmonic envelope, as the projection of the intersection of the surface $F = 0$ and the surface $\frac{\partial^u F}{\partial t^u} = 0$ where u is any positive integer. These harmonic envelopes suggest some natural conjectures. That is, in Figure 12b, the curve $\mathcal{F}_2$ looks like a trochoid; and sifting through the equations yields the following result where we use the word *harmonic* in the sense of obtaining similar patterns at deeper levels.

Theorem 2 (Harmonic Envelopes) *Let n be a positive integer. Let F be the function in (8). The curve $\mathcal{F}_{2n}$ is a trochoid whose parametric representation is*

$$\left(\frac{(p^{2n} - (q - p)^{2n}) \cos(pt) + ((q - p)^{2n} - q^{2n}) \cos(qt)}{p^{2n} - q^{2n}}, \right.$$
$$\left. \frac{(p^{2n} - (q - p)^{2n}) \sin(pt) + ((q - p)^{2n} - q^{2n}) \sin(qt)}{p^{2n} - q^{2n}} \right). \tag{11}$$

To generate (11) note that $\frac{\partial^{2n} F}{\partial t^{2n}} = 0$ is equivalent to

$$x(p^{2n} \sin(pt) - q^{2n} \sin(qt)) + y(q^{2n} \cos(qt) - p^{2n} \cos(pt))$$
$$+ (q-p)^{2n} \sin(qt - pt) = 0.$$

From $F = 0$, we have $x = (\sin(pt - qt) + y(\cos(pt) - \cos(qt)))/(\sin(pt) - \sin(qt))$. Substituting this expression into $F_t = 0$ along with straightforward simplification, we get the desired equation for y. The formula for x is similarly derived; (11) is a special case of (4).

Figure 12d shows the trochoid $\mathcal{F}_4$. Odd-order derivatives fail to yield trochoids however; for example Figure 12c gives $\mathcal{F}_3$. Experimenting with other combinations give interesting curves, as in Figures 12e and 12f.

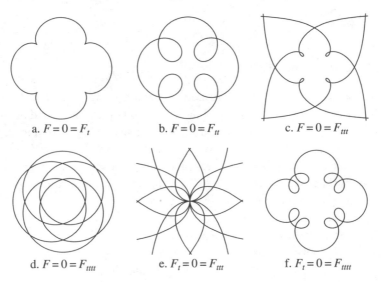

a. $F = 0 = F_t$ b. $F = 0 = F_{tt}$ c. $F = 0 = F_{ttt}$

d. $F = 0 = F_{tttt}$ e. $F_t = 0 = F_{ttt}$ f. $F_t = 0 = F_{tttt}$

Figure 12. Harmonic envelopes for $p = 5, q = 1$

Interpreting the second harmonic

To understand the significance of the second harmonic as a trochoid, let us recall from linear algebra the formula for the distance between a point $X = (x, y)$ and a line $\mathcal{L}$ through the points A and B. The projection vector $\mathbf{P}$ of the vector $X - A$ onto the line through the origin in the $B - A$ direction is

$$\mathbf{P} = \frac{(X - A) \cdot (B - A)}{(B - A) \cdot (B - A)} (B - A).$$

Hence the Euclidean distance from X to $\mathcal{L}$ is the distance between $X - A$ and $\mathbf{P}$, which is the square root of the dot product of $(X - A) - \mathbf{P}$ with itself, which upon simplification is

$$H(x, y, t) = \sqrt{(X - A) \cdot (X - A) - \frac{((X - A) \cdot (B - A))^2}{(B - A) \cdot (B - A)}}, \qquad (12)$$

where we identify A and B with the position of the two walkers holding the ends of the bungee cord, namely, $A = (\cos pt, \sin pt)$ and $B = (\cos qt, \sin qt)$. We claim that F is a shadow of H in the sense that except when A and B occupy the same location, when $H(x, y, t)$ is far from 0 then so is $F(x, y, t)$. That is, for almost all t values, the solution set in the xy-plane for $F(x, y, t) = 0$ is the same as that for $H(x, y, t) = 0$. The reason for the disparity is that when the walkers meet at time t with $(q - p)t$ being an integer multiple of π, then the solution set to $F(x, y, t) = 0$ is the entire xy-plane whereas the solution set to $H(x, y, t) = 0$ is the empty set. Figure 13 shows the typical relation between F and H; in particular, it compares the functions $F(0, .85, t)$ (the solid curve) and $H(0, .85, t)$ (the dotted curve) for $p = 2$, $q = 1$; the roots of the two curves agree except at $t = 0$ and $t = 2\pi$, the place where the walkers meet. The point $(0, .85)$ is on the bungee cord at $t \approx 1.14$ and $t \approx 3.89$. The reader can verify that the trochoid of the first harmonic $\mathcal{F}_1$ appears to be identical to that curve given by the solution set to

$$H(x, y, t) = 0 \quad \text{and} \quad H_t(x, y, t) = 0.$$

Thus, F can be interpreted as the signed distance from the bungee cord to (x, y) at time t. When $F(x, y, t) > 0$ the point (x, y) is above the bungee cord at time t and when $F(x, y, t) < 0$ the point (x, y) is below the cord at time t. The solution to the system $\partial^2 F / \partial t^2 = F_{tt} = 0$ includes the points at which $F(x, y, t)$ is changing most rapidly with respect to t. That is, the solution set to $F = 0 = F_{tt}$ includes those points for which the bungee cords are sweeping over the point (x, y) at time t most rapidly. To use a baseball anology, if we think of the point (x, y, t) as a baseball at (x, y) being hit by a bat at time t and $F_t(x, y, t)$ as the speed of the bat at time t, then the condition $F = 0 = F_{tt}$ translates into hitting only those baseballs that can be hit as hard as is possible, so the second harmonic characterizes those places of optimal impact.

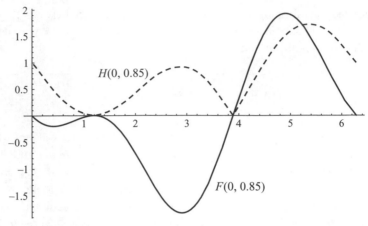

Figure 13. Distance from bungee cord to $(0, 0.85)$, where $p = 2, q = 1$

Some geometry of the higher harmonics

The significance of the higher harmonic trochoids is unclear to me. However, we can give some geometrical insight into the warping and rotating influence of the uth derivative acting on the original family of lines. This involves a fractional derivative. For any real number u, let the curve $\mathcal{F}_u$, the uth harmonic envelope, be defined as the xy-projection of the solution set of $\{F = 0, \frac{\partial^u F}{\partial t^u} = 0\}$. With this operator we can see how the epicycloid $\mathcal{F}_1$ in Figure 12a morphs into the trochoid $\mathcal{F}_2$ of Figure 12b as u increases from 1 to 2. As will be shown, the derivative metamorphosis of $\mathcal{F}_1$ becoming $\mathcal{F}_2$ encounters a stage $\mathcal{F}_{1.35}$ which is similar to the stage $\mathcal{F}_3$ in a metamorphosis of $\mathcal{F}_2$ into $\mathcal{F}_4$.

Fractional derivatives (see [59] and [66]) date back to 1867. Their definition uses the gamma function: $\Gamma(\alpha) = \int_0^\infty x^{\alpha-1} e^{-x} dx$, where $\alpha > 0$. Extending this to all real numbers by using the recursion relation $\Gamma(\alpha) = \Gamma(\alpha + 1)/\alpha$ gives the graph in Figure 14a. Then the intimidating definition due to Grünwald (1867) of the uth fractional derivative expanded about a where u and a are any real numbers is

$$\frac{d^u f(x)}{[d(x-a)]^u} = \lim_{N \to \infty} \left\{ \frac{[\frac{x-a}{N}]^{-u}}{\Gamma(-u)} \sum_{j=0}^{N-1} \frac{\Gamma(j-u)}{\Gamma(j+1)} f\left(x - j\frac{x-a}{N}\right) \right\}. \quad (13)$$

With a few appropriate tricks, we can find $\mathcal{F}_u$. First of all, we take $a = 0$. Two familiar properties of the ordinary derivative operator carry over

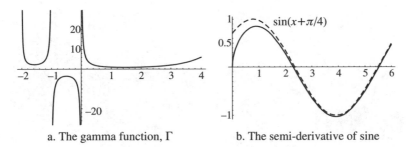

a. The gamma function, Γ   b. The semi-derivative of sine

Figure 14. The Γ function and the semi-derivative

as properties of the fractional derivative. For any functions f and g and constants α and β we have

$$\text{Linearity: } \frac{d^u(\alpha f(x) + \beta g(x))}{dx^u} = \alpha \frac{d^u f(x)}{dx^u} + \beta \frac{d^u g(x)}{dx^u}. \tag{14}$$

$$\text{A Primitive Chain Rule: } \frac{d^u f(\alpha x)}{dx^u} = \alpha^u \frac{d^u f(\alpha x)}{d(\alpha x)^u} \quad \text{where } \alpha > 0. \tag{15}$$

Some familiar properties of the ordinary derivative are hopelessly absent. In particular, it turns out that the uth derivative of a periodic function fails in general to be periodic. To see this, let us consider the semi-derivative, where $u = 1/2$. Applying the definition of Courant and Hilbert [**59**, p. 50] to the sine function gives

$$\frac{d^{\frac{1}{2}} \sin x}{dx^{\frac{1}{2}}} = \sqrt{2} \int_0^{\sqrt{\frac{2x}{\pi}}} \cos\left(x - \frac{\pi}{2}t^2\right) dt. \tag{16}$$

Fresnel integrals of this kind are found in the standard reference [**2**, p. 300]. The solid curve of Figure 14b depicts the graph of the semiderivative of $\sin x$ on $[0, 2\pi]$. As x increases, the semi-derivative of $\sin x$ converges to $\sin(x + \pi/4)$, the dashed curve of Figure 14b. In general, for large x it can be shown [**59**, p. 110] that

$$\frac{d^u \sin x}{dx^u} = \sin\left(x + \frac{\pi}{2}u\right) + \frac{x^{-1-u}}{\Gamma(-u)} - \frac{x^{-3-u}}{\Gamma(-u - 2)} + \cdots$$

and

$$\frac{d^u \cos x}{dx^u} = \cos\left(x + \frac{\pi}{2}u\right) + \frac{x^{-2-u}}{\Gamma(-u - 1)} - \frac{x^{-4-u}}{\Gamma(-u - 3)} + \cdots.$$

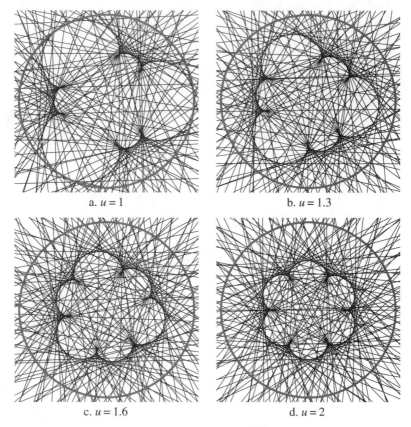

a. $u = 1$ b. $u = 1.3$

c. $u = 1.6$ d. $u = 2$

Figure 15. Envelopes of the family of lines, $\frac{\partial^u F}{\partial t^u} = 0$ with $p = 4, q = 1$

Thus, for $u > 0$ and for x sufficiently large,

$$\frac{d^u \sin x}{dx^u} \approx \sin\left(x + \frac{\pi}{2}u\right) \quad \text{and} \quad \frac{d^u \cos x}{dx^u} \approx \cos\left(x + \frac{\pi}{2}u\right). \quad (17)$$

By (14), (15), and (17), for $p \leq q$, $\frac{\partial^u F}{\partial t^u} = 0$ is asymptotically the same as

$$x\left(p^u \sin\left(pt + \frac{\pi u}{2}\right) - q^u \sin\left(qt + \frac{\pi u}{2}\right)\right)$$
$$+ y\left(q^u \cos\left(qt + \frac{\pi u}{2}\right) - p^u \cos\left(pt + \frac{\pi u}{2}\right)\right) \quad (18)$$
$$+ (q - p)^u \sin\left((q - p)t + \frac{\pi u}{2}\right) = 0.$$

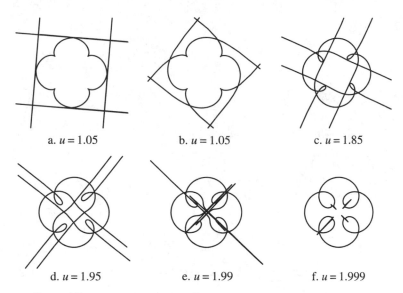

a. $u = 1.05$ b. $u = 1.05$ c. $u = 1.85$

d. $u = 1.95$ e. $u = 1.99$ f. $u = 1.999$

Figure 16. An epicycloid morphing into the trochoid $\mathcal{F}_2$, $p = 5$, $q = 1$

If $p > q$, the last term on the left in (18) can be replaced with

$$-(p - q)^u \sin((p - q)t + \pi u/2).$$

For each u, the family of lines $\partial^u F/\partial t^u = 0$ parametrized by t has an envelope. Figure 15, where $p = 4$ and $q = 1$, illustrates this for the u values 1, 1.3, 1.6, and 2, suggesting that the uth partial derivative acting on the family of lines given by $F = 0$ tends to constrict and rotate them into another family of lines. Against the unit circle shown in gray, the envelope shrinks and rotates clockwise as u increases from 1 to 2.

Let us define $\mathcal{G}_u$ to be the projection onto the xy-plane of the intersection of the surface $F = 0$ with the surface given by (18). As t gets large, $\mathcal{F}_u$ approaches $\mathcal{G}_u$. Thus we can view $\mathcal{F}_u$ as being asymptotically periodic.

The graphs of these asymptotic harmonic envelopes, as in Figure 16, show how the epicycloid of Figure 12a changes with the fractional derivative into the trochoid of Figure 12b. To describe this transformation, we imagine a surgical team of four simultaneously snipping Figure 12-1 at the four points furthest from the origin, so creating eight thrashing strands which explode outwards to the point at infinity somewhat like the severed ends of a piano string erupting under great tension; each surgeon grapples with the two strands created, wrestling their communal patient clockwise, causing the curve to cross at its cusps so forming loops as they push the strands back into

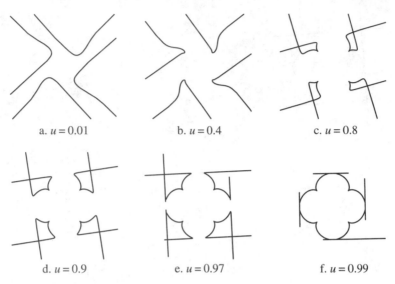

a. $u = 0.01$ b. $u = 0.4$ c. $u = 0.8$

d. $u = 0.9$ e. $u = 0.97$ f. $u = 0.99$

Figure 17. An epicycloid emerging from nothing, $p = 5, q = 1$

themselves, and ultimately meld the ends at the four points nearest the origin. Figure 17 shows how the trochoid emerges from nothing into an epicycloid by examining the fractional derivative harmonic envelopes as u ranges from 0 to 1.

A generalization

Just as the wheel model of the trochoid generalizes to multiple wheels in Farris [**29**], so the bungee cord model generalizes. That is, if $\sum_{i=1}^{n} \lambda_i = 1$ and a_i are rational numbers, then any curve of the form

$$\left(\sum_{i=1}^{n} \lambda_i \cos(a_i t), \sum_{i=1}^{n} \lambda_i \sin(a_i t) \right)$$

is a generalized trochoid. Such a generalized trochoid is generated by following the path of a hub attached to n bungee cords whose other ends are held by n walkers where walker i moves around the unit circle with speed a_i. Figure 18 depicts the path of a hub attached to three cords whose ends are held by walkers A, B, and C positioned at $(\cos t, \sin t)$, $(\cos 5t, \sin 5t)$ and $(\cos 17t, \sin 17t)$, and $\lambda_1 = 1/2$, $\lambda_2 = 1/3$, and $\lambda_3 = 1/6$, respectively.

The primary thrust of this chapter has been to generate appealing, periodic, trochoidal-like curves using the idea of an envelope. In the next chapter we

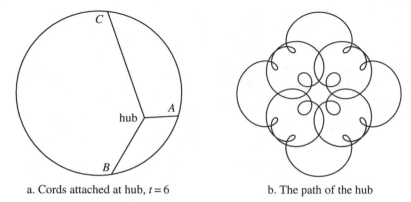

a. Cords attached at hub, $t = 6$ b. The path of the hub

Figure 18. A three-walker trochoid

imagine that the earth rotates about the sun in a circle and that the moon rotates about the earth in a circle. Hence the moon moves about the sun in trochoidal fashion. We use this observation to help H.G. Wells navigate his spaceship from the moon to the earth. In Chapter XIII we return again to the envelope, using it to analyze the rotating beacon problem that appears in the related rates section of a typical calculus text, and wind up generating an exotic curve. Spinning motion creates lovely natural patterns if only we have the eyes to see them.

Exercises

1. Verify that the ellipses $(x, y) = (\cos t, a \sin t)$ are the trochoids $T(1, -1, (a + 1)/2)$.

2. Verify that the polar flowers $r = \cos(m\theta/n)$ where m and n are integers arc the trochoids $T(n + m, n - m, 1/2)$.

3. Verify that the polar cardioid $r = 1 + a \cos\theta$, $a > 0$, is the translated trochoid $(a + 2)/2\, T(1, 2, a/(a + 2)) + (a/2, 0)$.

4. Verify that the epicycloid and hypocycloid of p cusps are $T(p + 1, 1, 1/(p + 2))$ for $p \geq 0$ and $T(p - 1, -1, 1/(2 - p))$ for $p \geq 3$, respectively.

5. Characterize those trochoids whose loops fail to intersect (with other loops). Some trochoids of p non-intersecting inner and outer loops are $T(p + 1, 1, (1 + 2p)(p^2 + 2p))$ and $T(p - 1, -1, (1 - 2p)(p^2 - 2p))$ for $p \geq 3$.

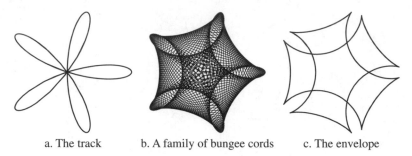

a. The track   b. A family of bungee cords   c. The envelope

Figure 19. An envelope on the polar flower $r = \cos(5\theta)$

6. Find a bungee cord characterization of the general trochoid. Find other familiar or unusual curves that can be generated from elementary function pairs (f, g).

7. Classify the curves in the xy-plane as given by the solution set to the system $\partial^u F / \partial t^u = \partial^v F / \partial t^v$, where u and v are positive integers. Is the case for $u = 1$ and $v = 4$ as given in Figure 12 the most exotic of these curves?

8. Each family of lines given by $\partial^u F / \partial t^u = 0$ as indexed by t has an envelope where u is a real number, as exemplified in Figure 15. Some of these are epicycloids and hypocycloids. How are these envelopes related (as u changes)?

9. Determine whether the first harmonic, $\mathcal{F}_1$, is the same as the projection of intersection of the surfaces $H = 0$ and $H_t = 0$, for H as given in (9). Contrast the second harmonic, $\mathcal{F}_2$ with the curve as defined by $H = 0 = H_{tt}$. Try this question: does $\lim_{u \to 0} \mathcal{F}_u$ exist? (Refer back to Figure 17a.)

10. In this chapter we allowed two walkers to proceed about the unit circle connected by a bungee cord. Now allow them to proceed about any trochoid. In particular, try the polar curve $r = \cos p\theta$, following Maurer [**54**], where p is a positive integer with walkers A and B positioned on the track at the polar angles $\theta + \delta$ and θ, respectively, for a fixed δ, while θ ranges from 0 to 2π so that the radial positions of A and B are $\cos p(\theta + \delta)$ and $\cos p\theta$, respectively. What kinds of paths are traced out by tacks embedded in such a bungee cord? Figure 15 shows the cords and the envelope corresponding to $p = 5$ and $\delta = 97\pi/180$. Are any of these familiar curves?

preamble xi

> The world turned upside downward.
>
> Robert Burton (1577–1640),
> *The Anatomy of Melancholy*

A new member of a department is often given less than glamorous assignments. When I arrived on sabbatical in Dar es Salaam, Tanzania, the mathematics chairwoman welcomed me as seminar coordinator, as well as instructor for two classes. Over the year, I put together a small slate of speakers. They were hard to come by. Few mathematicians traveled through East Africa.

As I thought about the lack of speakers while also wanting to visit Botswana where I had taught during a previous sabbatical, I realized that other East African countries suffered a similar plight. Their math departments lacked visiting speakers too.

Why not organize a lecture tour for myself? In America, I never would be so brash. But with a dearth of mathematical travelers in Africa, why not?

Soon I had invitations from the national universities of Zambia, Zimbabwe, and Botswana. At each place, I was warmly received, cordially heard, and asked to greet specific colleagues at the next venue. At each place, I was the only outside speaker they had all year.

However, getting there was an adventure. My Swahili skills were such that I understood the train ride to Zambia would take fourteen hours. When the train pulled out on time with me onboard, I congratulated myself on a well-laid plan. But after a day on the train, I realized that the ride contained fourteen stops and lasted forty-two hours. In Zambia's capital, I asked about the loud, frequent car backfirings. "No," I was assured, "that's rifle fire." That night, in a diplomat's compound, I turned off what I thought were the lights. Instead, the switch called out a contingent of U.S. marines who in moments were at the door with weapons ready. On the night train south to Zimbabwe, I lay diagonally on a small upper berth while the back and forth amplitude of the coach over poorly maintained tracks caused my legs and head to rise clear off the mat all night long, like some marathon love-making to a mattress. When I crossed from Zimbabwe to Botswana, a police sergeant interrogated me for four hours, insisting I was a smuggler of stolen vehicles. He demanded that I come up with a story better than that of being an American mathematician giving talks at East African universities, a story too incredible to believe.

This next chapter concerns a much more incredible story, an impromptu trip to the moon using a gravity cloaking material. We shall see if it makes any mathematical sense.

CHAPTER XI

Retrieving H. G. Wells from the Moon

H. G. Wells published *The First Men in the Moon* in 1901, a science fiction classic about two Englishmen who build a large, glass sphere, coat it in a gravity-blocking substance called cavorite, and then fly off to the moon. Wells never names his spaceship; so let's give it a name, Cavor, since we will refer to the spacecraft frequently. Within Cavor's hull of thick glass, covered outside by cavorite-coated steel shutters, are levers whereby the shutters may be opened and closed. Imagine that the shutters can be operated so that the gravitational effect of any specific heavenly body either is brought to bear on the entire mass of the ship or has absolutely no influence on the motion of the ship. With these shutters, the scientist in the story claims that

> we shall be able to tack about in space just as we wish. Get attracted by this and that [**95**, p. 408].

At lift-off from the earth,

> There came a little jerk, a noise like champagne being uncorked in another room, and a faint whistling sound. For just one instant I had a sense of enormous tension, a transient conviction that my feet were pressing downward with a force of countless tons [**95**, p. 408].

Wells' scientist explains,

> We are flying away from the earth at a tangent, and as the moon is near her third quarter, we are going somewhere towards her [**95**, p. 415].

After a number of days and a madcap opening and closing of the shutters, Cavor lands on the moon:

> There came a jar and then we were rolling over and over, bumping against the glass and against the big bale of our luggage, and clutching

a. From the first edition [94].

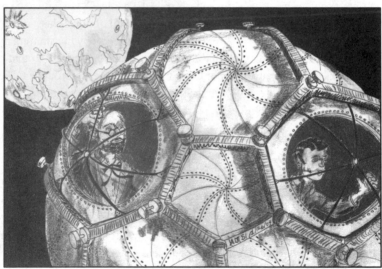

b. A sketch of the cavorite sphere.

Figure 1. H. G. Wells' spaceship

at each other, and outside some white substance splashed as if we were rolling down a slope of snow [**95**, p. 420].

After a series of adventures on the moon, Cavor returns to earth through a sequence of opening and closing of the cavorite shutters. Although it took "some weeks of earthly time" [**95**, p. 487] for the return trip, Cavor fortuitously plops into the ocean just off the coast of England. As the narrator later reflects on his strategy in returning to earth,

> Had I known then, as I know now, the mathematical chances there were against me, I doubt if I should have troubled even to make any attempt [at returning home] [**95**, p. 487].

We model Cavor's flight from the earth to the moon and back again. We ignore the question about the existence of a cavorite-like substance and make a few simplifying assumptions. The strategy for going from the earth to the moon is fairly simple. Returning from the moon is a greater challenge. In particular, we are interested in Wells' prediction that the return flight lasts several weeks. Can we do better using only on-off gravity as the propulsion for Cavor?

Some assumptions

We assume that the moon rotates in a circle about the earth and that the center of the circle is the center of the earth, even though the actual earth-moon system's barycenter, the center of mass of the system, is about 1707 km below the earth's surface. Table 1 gives some pertinent physical data. For ease of writing a subsequent parametrization, we take the moon's period about the earth as the sidereal period (27.3 days), that is, with respect to the stars, rather than the synodic period (29.5 days), which is the period with respect to the sun. Thus, Q and q have the same value. But the formulas we derive shall handle the general case when a satellite's period about its host planet and the satellite's period about its own axis are different. We say that the spaceship *shutters on* or *shutters off* a heavenly body's gravitational pull if its shutters are open or closed towards that body, respectively.

From the earth to the moon, a first model

In this section, we assume that the earth and the moon are the only bodies in the heavens. Let's launch Cavor from the earth's equator, perhaps near Mount Kilimanjaro. Once the astronauts are aboard and the earth is shuttered off, Cavor flies off along the tangent line to the earth's equator, as shown in Figure 2. Cavor's velocity is $2\pi r_e$ km/day, which is the product of the earth's radius and its angular velocity. At that rate, Cavor reaches the moon's orbit in about 9.5 days. (See Exercise (1) for determining a launch time so that the Cavor-moon rendezvous occurs in 9.5 days.) If the launch site is near

Table 1. Some physical constants

earth's radius	r_e	6400	km
moon's radius	r_m	1738	km
distance from earth to moon	r_d	384000	km
distance from earth to sun	r_s	1.50×10^8	km
gravitational acceleration at earth's surface	g	9.8	m/sec^2
moon's (sidereal) period about the earth	Q	27.3	earth days
moon's period about its axis	q	27.3	earth days
gravitational constant for earth	k	gr_e^2	

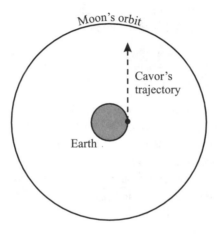

Figure 2. Going from the earth to the moon

London at 52° latitude, then the effective radial distance of the site from the earth's axis is $r_e \cos(52°)$ which means that reaching lunar orbit takes about 15.5 days. In Wells' story, as the astronauts near the moon they manipulate levers so as to achieve a soft lunar landing. But let us be content with just reaching the moon's orbit.

From the earth to the moon, a second model

Cavor's trip from the earth to the moon can be shortened if we use the sun. Assume that the earth orbits the sun in a circle, and that the sun, earth and moon are all in the same plane. With respect to the sun, the earth is at position

$$E(t) = r_s \left(\cos \frac{2\pi t}{365}, \sin \frac{2\pi t}{365} \right),$$

where t is in days. Also with respect to the sun, the unlaunched spacecraft is at position

$$S(t) = E(t) + r_e \big(\cos(2\pi t), \sin(2\pi t) \big).$$

At time p, launch Cavor by shuttering off all three heavenly bodies. Cavor will fly off along the tangent to its path as given by $S(t)$ with velocity $S'(p)$. Hence, Cavor's route will be the line

$$C(t, p) = S(p) + S'(p)(t - p),$$

for $t \geq p$. (For $t < p$, define $C(t, p) = S(t)$.) Once launched, Cavor's distance from earth's center is given by the magnitude of $C(t, p) - E(t)$, for $t \geq p$.

To understand the dynamics of these equations better, let us temporarily change the values of some parameters since the disparity between r_s and r_d of Table 1 tends to obscure what we wish to see. For the moment, let $r_s = 15$, $r_d = 1$, $r_e = 0.1$. In Figure 3, the dashed curve represents a portion of the orbit of the earth about the sun. At $t = 0$, Cavor is on the earth at position

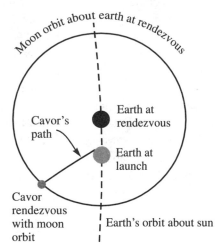

Figure 3. Cavor's path for an illustrative sun-earth-moon system: earth launch at time $p = 0.4$ days, moon rendezvous at time $t \approx 2$ days

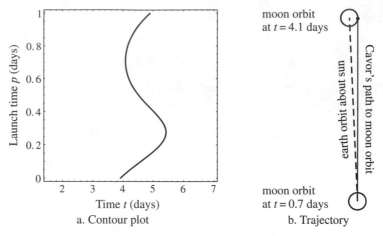

moon orbit
at $t = 4.1$ days

Cavor's path to moon orbit

earth orbit about sun

moon orbit
at $t = 0.7$ days

a. Contour plot

b. Trajectory

Figure 4. Cavor's flight to the moon for our sun-earth-moon system: earth launch at time $p = 0.7$ days, moon rendezvous at time $t \approx 4.1$ days

$(15.1, 0)$ with the sun at the origin. At $t = 0.4$ days, Cavor is launched, and flies off along the line as given by $C(t, 0.4)$. At $t = 2$ days, Cavor is one unit from the center of the earth, which means that Cavor's total flight lasted about 1.6 days. No matter when Cavor leaves the earth, the flight time is almost invariant for these values of $r_s, r_d,$ and r_e.

However, Cavor's flight time for our sun-earth-moon system is not invariant with respect to launch time. The easiest way to see this is to look at a contour plot of the function giving Cavor's distance from the earth with respect to time t and launch time p, as shown in Figure 4a; the single contour corresponds with Cavor being distance r_d from the earth. That is, for each ordered pair (t, p) on the curve, if Cavor is launched at time p, it will reach the moon's orbit at time t; for Cavor to rendezvous with the moon at time t, the astronauts must wait for an appropriate earth-moon orientation at launch so that the moon meets Cavor at time t. For launch time $p = 0$, the rendezvous occurs at about four days. However, the least flight time corresponds to launch time $p \approx 0.7$ days, with moon rendezvous at $t \approx 4.1$ days, which means that the flight time is about 3.4 days! Figure 4b shows these dynamics drawn to scale for our sun-earth-moon system.

From the moon to the earth, a first model

To bring Cavor home from the moon, we first assume that the earth and the moon are the only bodies in the heavens. That is, assume that the earth is fixed in space, and that the moon rotates about it.

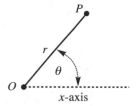

Figure 5. A particle P in a gravitational force field radiating from O

To find a way home we need Newton's planetary motion result from (12) in Chapter IV or from any calculus text, such as [**27**, pp. 827–30]: the path of a particle at position P about a mass centered at the origin O is given by the polar ellipse

$$r = \frac{1}{a\cos(\theta + \phi) + k/h^2}, \tag{1}$$

where r is the distance between P and O, θ is the angle between the positive x-axis and P, a is a constant, ϕ is a constant phase angle, and k and h are constants. Figure 5 shows the relation between r, θ, and P. In particular, k is the gravitational constant as determined by the mass at O; if this mass is the earth, then the earth induces an acceleration of $f(r) = -k/r^2$ on the particle at P when it is r units from O, with $k = gr_e^2$. The constant of angular momentum h is given by $h = r^2\frac{d\theta}{dt}$, where t is time.

In a lunar launch, imagine that we are free to place Cavor at any point P on the lunar equator which faces the earth. At the time of launch, we orient the coordinate system so that the earth is at the origin and P is on the x-axis. Let ω be the angle of the launch site on the moon; that is, ω is the angle between the ray at the moon's center M in the positive x direction and the ray from M through P, $\pi/2 \le \omega \le 3\pi/2$, as shown in Figure 6. Our launch strategy is to shutter off the moon and shutter on the earth simultaneously.

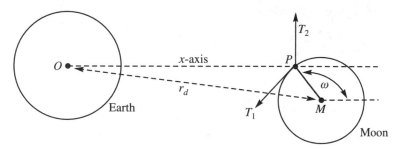

Figure 6. Orientation at moment of lunar launch, $\theta = 0$

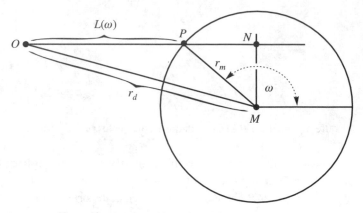

Figure 7. Distance of launch site from earth's center

We will find a and ϕ in terms of ω. The direction in which Cavor proceeds is the vector sum of two tangent directions: $\mathbf{T}_1$ to the moon at P and $\mathbf{T}_2$ to P's orbit about the earth. To find $\mathbf{T}_1$, we see that the ray from the moon's center M through P has direction $(\cos \omega, \sin \omega)$, so $\mathbf{T}_1$'s direction is $(-\sin \omega, \cos \omega)$. Therefore, as a function of ω,

$$\mathbf{T}_1(\omega) = \frac{2\pi r_m}{q}(-\sin \omega, \cos \omega).$$

To find $\mathbf{T}_2$, we see that the distance of P from O is

$$L(\omega) = \sqrt{r_d^2 - r_m^2 \sin^2 \omega} + r_m \cos \omega. \qquad (2)$$

To see this, drop a perpendicular from M to the line through $\overline{OP}$, as indicated in Figure 7. Let N be the point of intersection of these two lines. The length of $\overline{MN}$ is $r_m \sin \omega$, so the length of $\overline{ON}$ is $\sqrt{r_d^2 - r_m^2 \sin^2 \omega}$. The length of $\overline{PN}$ is $r_m \cos \omega$, a negative quantity for $\pi/2 \leq \omega \leq 3\pi/2$. The sum of these two quantities gives (2).

Therefore, as a function of ω,

$$\mathbf{T}_2(\omega) = \frac{2\pi L(\omega)}{Q}(0, 1).$$

Thus

$$\mathbf{T}_1(\omega) + \mathbf{T}_2(\omega) = \big(v_x(\omega), v_y(\omega)\big)$$

$$= \left(-\frac{2\pi r_m \sin \omega}{q}, \frac{2\pi r_m \cos \omega}{q} + \frac{2\pi L(\omega)}{Q}\right).$$

At launch, Cavor's initial angular velocity about the earth is $v_y(\omega)/L(\omega)$, being Cavor's velocity perpendicular to line $\overline{OP}$ divided by the distance of Cavor from the earth. Once launched from the moon (with the moon's gravity shuttered off and the earth's gravity shuttered on), Cavor's path follows the conic (4). The constant k is known, whereas h is now a function of ω, but once Cavor is launched from the moon, h remains fixed as long as earth's gravity is continuously applied. This means that the constant of angular momentum for Cavor about the earth with respect to ω is

$$h(\omega) = L^2(\omega)\frac{v_y(\omega)}{L(\omega)} = L(\omega)v_y(\omega).$$

Rewriting (4), we have $1/r = a\cos(\theta + \phi) + k/h^2(\omega)$. Differentiating with respect to t, gives

$$-\frac{1}{r^2}\frac{dr}{dt} = -a\sin(\theta + \phi)\frac{d\theta}{dt}.$$

At $\theta = 0$, we know that $r = L(\omega)$, $\frac{dr}{dt} = v_x(\omega)$, and $\frac{d\theta}{dt} = h(\omega)/L^2(\omega)$. Therefore, $a\sin(\phi) = v_x(\omega)/h(\omega)$. From (4) we also know that $a\cos(\phi) = 1/L(\omega) - k/h^2(\omega)$. These last two expressions yield

$$a^2 = \left(\frac{v_x(\omega)}{h(\omega)}\right)^2 + \left(\frac{1}{L(\omega)} - \frac{k}{h^2(\omega)}\right)^2.$$

Without loss of generality, we take a as negative, giving

$$a(\omega) = -\sqrt{\left(\frac{v_x(\omega)}{h(\omega)}\right)^2 + \left(\frac{1}{L(\omega)} - \frac{k}{h^2(\omega)}\right)^2}. \tag{3}$$

Using this value for $a(\omega)$, we can then solve $a(\omega)\sin\phi = v_x(\omega)/h(\omega)$ for ϕ, giving

$$\phi(\omega) = \sin^{-1}\left(\frac{v_x(\omega)}{a(\omega)h(\omega)}\right). \tag{4}$$

If we launch Cavor from the hidden face of the moon, we will not be able to shutter off the moon's gravity and shutter on the earth's gravity simultaneously, so $\phi(\omega)$ makes sense only when $\pi/2 \leq \omega \leq 3\pi/2$. Figure 8 is a graph of $\phi(\omega)$.

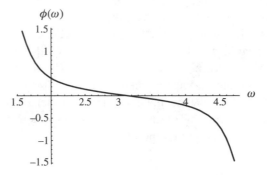

Figure 8. Graph of the phase angle $\phi(\omega)$ for $\pi/2 \le \omega \le 3\pi/2$

Since $a(\omega) < 0$ by (3), Cavor is nearest earth when $a(\omega)\cos(\theta + \phi(\omega)) + k/h^2(\omega)$ is greatest, which occurs when $\theta = \pi - \phi(\omega)$. Figure 9 is a graph of Cavor's closest approach to the earth with respect to its launch position given by ω.

As can be readily seen, the best place to launch Cavor appears to be at $\omega = \pi$, as the reader is invited to prove. However, that launch site fails to bring Cavor home. Figure 10 shows Cavor's trajectory as a solid ellipse which is almost the same as the dashed circle of the moon's orbit.

What to do?

Wells followed a long tradition in supposing that the moon was inhabited. The Pythagoreans were among the first to propose the idea. They imagined that lunar animals were fifteen times larger than terrestrial ones and that the beauty of lunar flowers far exceeded those on earth. Plutarch discussed moon men, and concluded that they must be lighter than earth men. In a

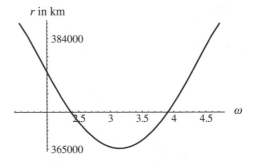

Figure 9. Graph of the nearest approach r to earth at $\theta = \pi - \phi(\omega)$ for $\pi/2 \le \omega \le 3\pi/2$

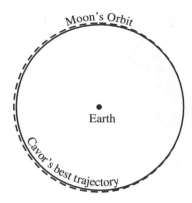

Figure 10. Failed attempt at returning home: Cavor's orbit when launched from the moon at $\omega = \pi$

1610 story of an imagined trip to the moon, Kepler described the moon as being

> porous and pierced through with hollows and continuous caves which are the chief protection of the inhabitants against heat and cold [**44**, p. 154].

For the features of the moon, he coined the word *selenographical* rather than using geographical, as the Greek word for the moon is selene, meaning white gleam. The first lunar map-maker, Johann Hevelius (1611–1687), decorated his charts with living seas and people whom he called Selenites.

Wells' Selenites were ant-like inhabitants of the depths of the moon. Figure 11a is a sketch of these creatures from his book's first edition. If the Selenites could give Cavor a running start before the astronauts shutter off the moon's gravity and shutter on earth's gravity, Cavor might make it home. Rather than thinking of a train moving in the direction of the moon's rotation across the lunar seas with Cavor as its payload, let us simply change the value of q, the moon's period about its axis. For example, for $q = .01\,Q$ and $\omega = 3\pi/4$, Cavor's nearest approach to the earth is about 110,000 km, over two-thirds of the way home. We're getting closer! Figure 11b shows Cavor's orbit.

The moon rotates about its axis once each 27.3 days, or at the rate of about 16.6 km per hour at the lunar surface. In order for q to equal $.01\,Q$, we must bring Cavor's speed on a lunar road up to 1644 km per hour before its launch.

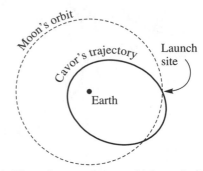

a. Selenites! [94]   b. The trajectory when $q = .01Q$, $\omega = 3\pi/4$

Figure 11. Hopeful progress

Unfortunately, the Selenites may not have the technology to achieve such a ground speed.

From the moon to the earth, a modified first model

Another idea to use so as to help Cavor home is the Tower of Babel method. Arthur C. Clarke used this idea in a sequel to his novel *2001* [21, pp. 28, 238–242], where he has a network of towers on the earth thirty-six thousand km high atop which spaceships may dock. So let's try this approach on the moon: build a tower and launch Cavor from that height. All of the formulas are in place. We simply change the value of r_m. But rather than stopping there, let's build a monorail 500 km high along the moon's equator and allow a train with Cavor as its payload to rush along the rail. Figure 12 shows Cavor's trajectory when $\omega = \pi$ and when the effective rotational rate is $q = .01Q$. On such a trajectory, Cavor comes to within about 30,000 km of the earth's surface, over 90% of the way home.

From the moon to the earth, a second model

When we allowed the sun in the second model to send Cavor from the earth to the moon, the time for Cavor's journey decreased. In like fashion, we allow the sun into our system in hopes of getting Cavor from the moon to the earth.

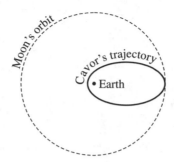

Figure 12. Almost there: the trajectory when $r_m = 2238$ km, $q = .01Q$, $\omega = \pi$

Here's our strategy. Let's first of all launch Cavor from the moon into an elliptical orbit about the earth as in Figure 10. From henceforth, ignore the moon. Since the earth rotates about the sun, Cavor will follow a kind of epicycloidal path as discussed in Chapter X. For simplicity, assume that Cavor orbits the earth in a circle of radius R with a period of q days. Let Z be the radius of the earth's orbit about the sun. Then the earth's position is given by

$$E(t) = Z\left(\cos\frac{2\pi t}{365}, \sin\frac{2\pi t}{365}\right),$$

while Cavor's position is given by

$$S(t) = E(t) + R\left(\cos\frac{2\pi t}{q}, \sin\frac{2\pi t}{q}\right),$$

where t is in days. At time p, shutter off both the earth and the sun so that Cavor moves along the tangent to the epicycloid with velocity $S'(p)$. Cavor's subsequent position at time t is given by

$$C(t, p) = S(p) + S'(p)(t - p), \tag{5}$$

where $t \geq p$. (For $t < p$, let $C(t, p) = S(t)$.)

For each time p, consider the intersection of the path of the earth about the sun and the tangent line to $S(t)$ at $t = p$ as it goes from inside earth's orbit to outside earth's orbit. To illustrate, take $Z = 15R$ and $q = 30$ days. In Figure 13, the dotted circular arcs represent the path of the earth moving counterclockwise around the sun and the solid epicycloids are the path of Cavor. As Figure 13a shows, there will be a time α when the intersection occurs when the earth has not yet passed the intersection point. That is, in Figure 13a, at time $\alpha = 9.5$ days, Cavor leaves the epicycloid, and travels

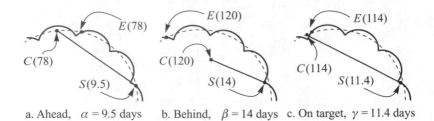

a. Ahead, $\alpha = 9.5$ days   b. Behind, $\beta = 14$ days   c. On target, $\gamma = 11.4$ days

Figure 13. The intermediate value theorem in action for an illustrative sun-earth-moon system

for 68.5 days, at which point it's at the end of the line segment in the figure, while the earth at 78 days is the big dot in the figure, so Cavor crosses earth's orbit ahead of the earth. As Figure 13b shows, there will be a time β, $\beta > \alpha$, when the intersection occurs when the earth has already passed the intersection point. When $\beta = 14$ days, at $t = 120$ the earth is at the big dot, and Cavor is well behind the earth, but has not yet crossed the earth's orbit. For times between α and β, intersection points continuously exist. Therefore, by the intermediate value theorem, there must be a time γ between α and β when Cavor, shuttering off both the earth and the sun, will rendezvous with the earth. (We won't worry about a soft landing.) Figure 13c shows such an occurrence, at $\gamma = 11.4$ days. At $t = 114$ days, the earth is at the big dot and Cavor has crashed into it.

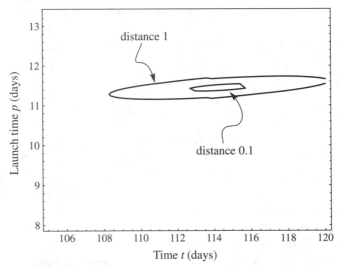

Figure 14. Finding a solution using a contour plot for the illustrative model

A more direct approach than the technique of Figure 13 is to use the contour plot technique of Figure 4a. Figure 14 shows a contour plot of the distance from Cavor to the earth at time t when launched at time p. For the illustrative system of Figure 4, recall that the distance between the earth and the moon is 1 unit, and the radius of the earth is 0.1 units. Figure 14 shows two contours corresponding to distance 1 and distance 0.1. That is, for any point (t, p) on the curve labeled distance 1, when Cavor is launched at time p, then at time t it will be 1 unit from earth. Any ordered pair (t, p) within the contour for distance 0.1 gives a Cavor-earth rendezvous. Figure 14 confirms that $(114, 11.4)$ is a solution.

We use this approach for our sun-earth-Cavor system, that is, when $Z = r_s$, $R = r_d$, and $q = 30$ days. Figure 15a shows the distance contour of 100,000 km. After zooming in, Figure 15b gives the contour of r_e and $(17.5, 12.5)$ fits comfortably within it, which means that if Cavor shutters off all heavenly bodies on day 12.5, then on day 17.5, Cavor will rendezvous with the earth, giving a return trip time of five days. (Exercise 8 repeats this calculation when $q = 27.3$ days.)

To help set this return time in context, in 1968, Apollo 8, the first manned craft to circle the moon, traveled from the earth to the moon in 3.5 days, and from the moon to the earth in 2.5 days. Of course, the Apollo craft used rocket power and gravity to navigate the heavens, where Cavor uses only on-off gravity.

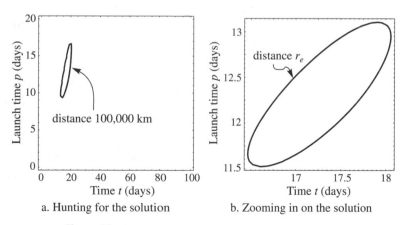

a. Hunting for the solution   b. Zooming in on the solution

Figure 15. Finding a solution trajectory via contour plots

Summary and stereotype

Wells was right to use the notion of tangent velocities as the underlying principle in guiding his spacecraft. He was also right to realize that the return trip is more complex than the outbound trip. However, with proper navigation, his projected return trip of several weeks can be shortened to about five days.

Finally, since I have used H.G. Wells' texts both in this chapter and in Chapter VI, I allow him a last word. At the end of his book, Wells gives an interesting stereotype of the mathematician. His lunar society had castes whose members were molded from an early age. Here is part of his description of the Selenite mathematician caste:

> If a Selenite is destined to be a mathematician his brain grows; his voice becomes a mere squeak for the stating of formulae; he seems deaf to all but properly enunciated problems. The faculty of laughter, save for the sudden discovery of some paradox, is lost to him; his deepest emotion is the evolution of a novel computation. And so he attains his end [**95**, p. 512].

Then as now mathematicians had image problems.

Exercises

1. Imagine that the earth and moon move in the same plane. Suppose at $t = 0$, the earth's center, Cavor, and the moon's center are collinear. When should Cavor be launched so as to rendezvous with the moon after a flight of 9.5 days?

2. For a harder problem than Exercise 1, assume that the moon orbits the earth in a plane inclined $5.9°$ with respect to the plane through the earth's equator. At $t = 0$, assume as before that the earth, Cavor, and the moon lie on the same line. When should Cavor be launched so as to rendezvous with the moon?

3. Examine Figure 4a to determine the launch time that maximizes Cavor's flight time to the moon. Adapt and solve the minimization problem with respect to Mars and its two moons.

4. Extend the graph of $\phi(\omega)$, Figure 7, from $0 \leq \omega \leq 2\pi$ even though Cavor can not be launched from the backside of the moon. What angle optimizes ϕ? Give an intuitive explanation as to why these angles are not $\pi/2$ and $3\pi/2$.

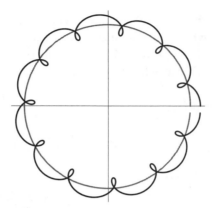

Figure 16. An epicycloid with loops

5. Prove that Cavor's best launch site on the moon is at $\omega = \pi$.

6. For $q = .01Q$ and $\omega = \pi$, plot Cavor's trajectory. Determine Cavor's nearest approach to the earth. For $\omega = \pi$, find the greatest value of q for which Cavor's nearest approach to the earth's center is precisely r_e. Furthermore, find the least value of q for which Cavor's trajectory lies along a hyperbola (or a parabola) rather than an ellipse, in which case Cavor will exit the earth-moon system.

7. Model the sun-earth-moon system so that the earth and moon rotate about their barycenter (see page 189) and so that the barycenter rotates about the sun in a circle. Find the shortest time for Cavor to travel from the moon to the earth. Contrast this value with the five day prediction on page 201.

8. Find the minimum time for Cavor to reach the earth assuming that $q = 27.3$ days rather than the 30 days of the example associated with Figure 15. Try the same problem using the moons of Mars, or your own sun-planet-spacecraft system.

9. The only case where Exercise 8 has no solution is when the epicycloid of Cavor about the sun has loops as in Figure 16, in which case the tangential velocity lines to the epicycloid always cross ahead of the earth, which means that for such a sun-earth-spacecraft system, Cavor won't be able to return to earth using the ideas of this section. In particular, show that for an epicycloid corresponding to $Z = 8$, $R = 1$, and $q = 30$ Cavor will not make it home.

10. To continue with Exercise 9, for $R = 1$ and $q = 30$, find the maximum value of Z for which Cavor is stranded in space. More generally, using the approach of this section characterize those triples (Z, R, q) for which Cavor will be able to return to earth.

preamble xii

Go, and catch a falling star.

John Donne (1572–1631), *A Song*

The Pseudospheres was the mathematics department graduate and faculty intramural team, participating with mixed success in various sporting events at the university. Being a member was a good way to get some exercise when not writing proofs.

Each season the position of coach rotated. The coach assembled the team, called practices, kept track of schedules, and made the rosters. Since being coach was work, we graduate students drew straws. One summer, I lost.

We weren't too bad and somehow we found ourselves playing for the softball championship. Our opponents were the graduate physical education team who had destroyed everyone they faced. As they took the field for warm-up practice, each player looked seven feet tall with broad shoulders, meaty biceps, and low Neanderthal brows. They peppered the field with line drives and soaring shots. It was easy to see why they hadn't lost a game.

Then it was our turn. As I hit the ball around for fielding practice, I couldn't help but feel proud of the team. I knew its weaknesses and strengths. A topologist played first base and a real analyst with a rocket arm played third. I had recruited a housemate for shortstop who had quick, sure hands, but was a dud at the plate. The center fielder was a graduate statistician who caught anything that was catchable. Right field was played by our weakest player. I kept myself out of the game for the first half.

As the game progressed, the P.E. team tested our defense, at first with confidence, knocking drives to center and left. "How could a mathematics team be any good?" they thought. Brute force should be enough to win easily. However, the left and center fielders were equal to the challenge. With the score tied halfway through the game, I knew that the P.E. grad strategy would change, that they would test the right side of the field. For this attack right-handed batters needed to swing a bit late. I put myself into right field. Sure enough, they focused on the new guy, me.

   A monster of a man came to the plate. He looked purposefully in my
direction, swinging his bat, his intention clear. I backed up. There was no
home run fence on this field. The ball arched from the pitcher's hand. I reacted
to the crack of the bat. It was a blast. I was all instinct. I raced backwards,
my face turned heavenward over my shoulder, my legs pounding the ground.
With my arm at full extension, the ball lodged in the webbing of the glove.
That catch broke the spirit of our opponents, and we went on to win, earning
a rare trophy for the display case of the math department.

CHAPTER XII

Playing Ball in a Space Station

Imagine playing ball inside a rotating space station. How will the artificial gravity affect the way a ball is thrown, kicked, or batted?

For instance, let's take the spaceship Discovery of Arthur C. Clarke's science fiction drama, *2001*. Figure 1 shows a sketch of the spaceship. The living quarters, a drum of diameter 35 feet, were inside a sphere of diameter 40 feet. The drum rotated once every ten seconds, thereby producing gravity on the drum floor equal to the moon's [**20**, p. 97]. That is, the centripetal acceleration, $\omega^2 R$, where ω is the rotation rate and R is the distance from the axis of rotation, gives an acceleration of 6.9 ft/sec^2.

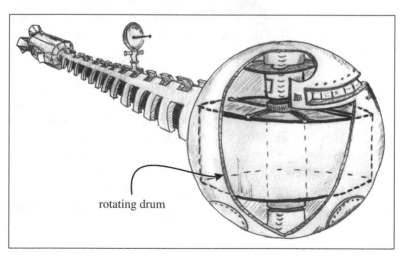

rotating drum

Figure 1. Cut-away of *2001*'s spaceship, Discovery

We shall see that throwing a ball on a space station can be significantly different from doing it on earth. For example, on a rotating space station it is possible to throw a ball so its trajectory has multiple loops.

Figure 2. The Little Prince on Asteroid B-612, [**24**, p. 15], courtesy of Harcourt

Before playing ball on Discovery, we first review ball dynamics on a planet. Depicted in Figure 2, Asteroid B-612 appears in the children's tale, *The Little Prince*. Since it is described as being "scarcely any larger than a house" [**24**, p. 16] and since we will compare its gravity with that of a rotating spacecraft, we assume that, like the drum in Discovery, the asteroid is 35 feet in diameter, and that its surface gravity is that of the moon. As such, the Little Prince's planet is extraordinarily dense, about 1.4 million grams/cm^3, the density of a white dwarf. This won't prevent us from playing ball on its surface.

Ball dynamics on Asteroid B-612

For most throws made on earth, a ball's trajectory is well-modeled by a parabola. However, if the speed of the ball is great enough, the ball will rise so high that the parabolic-trajectory-and-flat-earth model breaks down

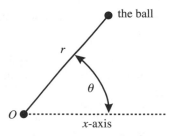

Figure 3. The polar variables

both because the downward acceleration is no longer constant and because the earth is circular. A much better model, ignoring air resistance, takes the ball as a small mass being attracted by the stationary mass of a planet. The trajectory of the ball then is a conic, as Kepler proposed and as Newton proved. A ball thrown so fast that it never returns to the planet's surface follows either a parabola or a hyperbola. Otherwise, it follows an elliptic path.

From any calculus text or from (12) in Chapter IV, the path of a thrown ball is given by the polar equation

$$\frac{1}{r} = a \cos(\theta + \phi) + \frac{k}{h^2}, \tag{1}$$

where a, ϕ, h, and k are constants.

To adapt (1) to model the path of a ball thrown on the asteroid, we position the origin O at the center of Asteroid B-612. Although the asteroid makes 44 revolutions per earth day, to make matters simple, we will assume that it doesn't rotate. Thus the path of a ball remains in a plane through O and the thrower, which means that we can perform our analysis in the xy-plane. The quantities used in (1) are

- t is time in seconds,

- r is the radial distance of the ball from O at time t,

- $R = 17.5$ feet, the radius of the asteroid,

- θ is the angle at O, made by the ball and the x-axis, as shown in Figure 3,

- $g = 6.9$ ft/sec^2 is the gravitational acceleration at the asteroid's surface,

- $k = gR^2$, a constant of proportionality, as follows from Newton's law of gravitation,

- $h = r^2 \frac{d\theta}{dt}$, the constant of angular momentum, as follows from the derivation of (1),

- a is a constant. In particular, if e is the eccentricity of the orbit, then $e = ah^2/k$ so that $a = ek/h^2$. We allow a, and thus e as well, to be positive or negative.

- ϕ is a constant phase angle with $-\pi/2 \le \phi \le \pi/2$. The reason we can use this convention is because $\cos(\theta + \pi) = -\cos\theta$ and because the sign of a can be positive or negative. Furthermore, $-\phi$ is the polar angle (direction) of the major axis of the orbit.

For the path of a ball, we adopt the convention that at time $t = 0$ the ball is thrown from the asteroid's surface at $A = (R, 0)$ and with initial velocity (p, q) as shown in Figure 4. For simplicity, we assume that the ball is thrown into the first quadrant, so that both p and q are non-negative. To find the path of the ball, all that we must do is to determine the constants a, ϕ, and the initial rotation rate $\frac{d\theta}{dt}\big|_{t=0}$. We will rewrite (1), replacing a, ϕ, h, and k with appropriate expressions in terms of p, q, g, and R.

To do so, let $x = r\cos\theta$ and $y = r\sin\theta$, where x, y, r, and θ are functions of t. Then, since $h = r^2\theta'$ and by use of the determinant,

$$\begin{vmatrix} x & x' \\ y & y' \end{vmatrix} = \begin{vmatrix} r\cos\theta & (r'\cos\theta - r\theta'\sin\theta) \\ r\sin\theta & (r'\sin\theta + r\theta'\cos\theta) \end{vmatrix} = r^2\theta' = h.$$

Furthermore, at $t = 0$,

$$\begin{vmatrix} x & x' \\ y & y' \end{vmatrix} = \begin{vmatrix} R & p \\ 0 & q \end{vmatrix} = Rq,$$

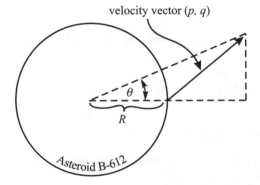

Figure 4. Throwing a ball with velocity (p, q) in the first quadrant

so $h = Rq$. Since $k = gR^2$, we have $h^2 = R^2q^2 = kq^2/g$, so $k/h^2 = g/q^2$. Since $k/h^2 = g/q^2$, (1) becomes

$$\frac{1}{r} = a\cos(\theta + \phi) + \frac{g}{q^2}. \tag{2}$$

The initial condition of throwing the ball from $(R, 0)$ means that when $t = 0, \theta = 0$ and $r = R$. Thus (2) gives

$$a\cos\phi = \frac{q^2 - gR}{q^2R}. \tag{3}$$

Differentiating (2) with respect to time gives

$$-\frac{1}{r^2}r' = -a\sin(\theta + \phi)\,\theta',$$

which at $t = 0$ gives

$$a\sin\phi = \frac{p}{qR}, \tag{4}$$

since $\frac{dr}{dt}|_{t=0} = p$, (because $x' = r'\cos\theta - r\theta'\sin\theta$ at $t = 0$ gives $x' = r'$). Combining (3) and (4) and taking the square root yields

$$a = \pm\frac{\sqrt{(pq)^2 + (q^2 - gR)^2}}{q^2R}. \tag{5}$$

By (3) and (4), it follows that

$$\phi = \tan^{-1}(pq/(q^2 - gR)). \tag{6}$$

Since $-\pi/2 \leq \phi \leq \pi/2$, then by (3), the sign in (5) is positive when $q^2 > gR$ and negative when $q^2 < gR$. (By (4), the sign is positive when $\phi = \pi/2$ and negative when $\phi = -\pi/2$.) Hence for $-\pi/2 < \phi < \pi/2$, the trajectory of a ball thrown into the first quadrant from $(R, 0)$ with velocity (p, q) is given by

$$\frac{1}{r} = \text{sign}(q^2 - gR)\frac{\sqrt{(pq)^2 + (q^2 - gR)^2}}{q^2R}$$

$$\times \cos\left(\theta + \tan^{-1}\left(\frac{pq}{q^2 - gR}\right)\right) + \frac{g}{q^2} \tag{7}$$

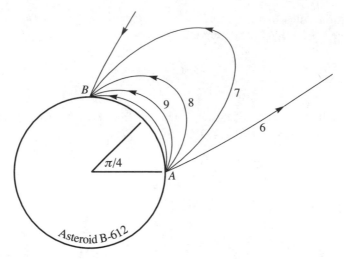

Figure 5. Throwing a ball from the equator to the north pole on Asteroid B-612

(and is $1/r = \pm p/(qR)\cos(\theta \pm \pi/2) + g/q^2$ for $\phi = \pm\pi/2$). For (7) to be the equation of an ellipse, its eccentricity e must be less than 1, so

$$\frac{\sqrt{(pq)^2 + (q^2 - gR)^2}}{q^2 R} < \frac{g}{q^2},$$

which simplifies to

$$p^2 + q^2 < 2gR. \tag{8}$$

So a ball thrown (in any direction!) at least as fast as $\sqrt{2gR} \approx 15.5$ ft/sec $\approx$ 10.6 mi/hr will escape Asteroid B-612. Even though its surface gravity is that of the moon, the Little Prince can almost run off his planet.

We use these results to see how to throw a ball from the equator at A to the north pole B, or to any other point on the surface. Rewrite (2) as

$$\frac{1}{r} = a\cos\phi\cos\theta - a\sin\phi\sin\theta + \frac{g}{q^2}.$$

From (3) and (4), we have

$$\frac{1}{r} = \frac{q^2 - gR}{q^2 R}\cos\theta - \frac{p}{qR}\sin\theta + \frac{g}{q^2}. \tag{9}$$

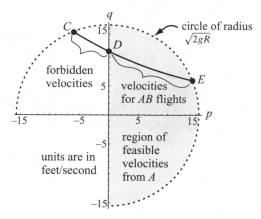

Figure 6. Locus of velocities (p, q) allowing throws from A to B

If we throw a ball from A to B, then the polar point $(R, \pi/2)$ must be on its trajectory. Substituting these values in (9) gives

$$p = \frac{gR - q^2}{q}. \tag{10}$$

This characterizes the initial velocity vectors (p, q) for which a ball thrown from the equator will reach the north pole. For example, when $q = 7$, $p \approx 10.25$. Figure 5 shows a few of these trajectories, labeled by the q component of the initial velocity. Furthermore, (10) can be written as $1 = pq/(gR - q^2)$; by (6), this means that ϕ is $-\pi/4$, so $\pi/4$ is the polar angle of the major axis, as indicated in Figure 5.

To put (10) in a more global perspective, recall that the p coordinate of the initial velocity of a ball thrown from A must be nonnegative. This observation, along with (8), means that the set of initial velocities (p, q) so that a ball thrown from A will return to the asteroid is the right half of the open disk of radius $\sqrt{2gR}$, centered at O, as indicated by the shaded region in Figure 6. The path as given by (10) is shown as a solid curve across the disk. The endpoints of this curve on the disk are labeled C and E. The velocities along the arc CD are forbidden, because they correspond to throwing a ball into the dirt. The locus of all velocity vectors allowing a ball thrown from A to B are those along the arc DE. The velocities along this arc near D correspond to trajectories with shorter flight times, while those near E correspond to longer flight times. This analysis of course can be generalized to any two points on a planet.

Ball dynamics on Discovery

Now we play ball aboard Discovery. We'll see that the trajectories can be very different from those shown in Figure 5. We ignore air resistance and assume that the mass of the space station has no effect on the motion of the ball. (See (6) in Chapter II for the gravity field within a hollow cylinder.) In this section we solve the special case of throwing a ball towards the drum's axis of rotation. The generalization to three dimensions will easily follow.

   The trajectory of a ball thrown towards the axis of rotation remains in a plane perpendicular to the axis, a plane that can be taken as the xy-plane. In this plane, the ball remains inside a circle of radius R, whose center is taken as the origin, O. As in the previous section, a ball with initial velocity (p, q) is thrown from position $A = (R, 0) = (17.5, 0)$. Such throws imply that $p \le 0$, since otherwise the ball will be thrown against the floor of the space station. Recall that the space station rotates at $\omega = \pi/5$ radians per second counterclockwise about its axis.

   To analyze the ball's motion we use two coordinate systems, one absolute and one apparent. The absolute frame is with respect to the background of the fixed stars, wherein the positive x-axis and y-axis always point to the same two stars. The apparent frame is with respect to the rotating spacecraft. Think of painting a reference dot D on the floor of the space station. With

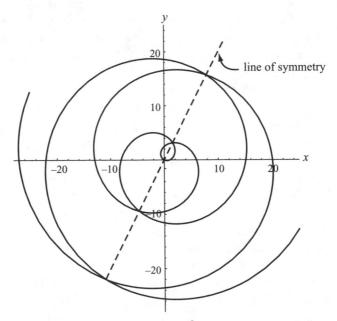

Figure 7. The spiral $(x, y) = t M(t) \begin{bmatrix} -2 \\ 1 \end{bmatrix}$, with symmetry line $y = 2x$

respect to D and O, the coordinates of each point in the disk of radius R remain unchanged in the apparent frame as it rotates.

With respect to the absolute frame, imagine that D is at $(R, 0)$ on the x-axis at $t = 0$. At that moment, since it is rotating counterclockwise about the central axis of the station, it experiences a rotational velocity of $(0, \omega R)$. The ball, initially at D, has the same rotational velocity. At $t = 0$ the ball is thrown with apparent velocity (p, q). Thus the ball's initial velocity with respect to the absolute frame is $(p, q + \omega R)$. With respect to the absolute frame, the ball's trajectory is the line

$$f(t) = \begin{bmatrix} R \\ 0 \end{bmatrix} + t \begin{bmatrix} p \\ q + \omega R \end{bmatrix} \tag{11}$$

from $t = 0$ until it hits the side of the space station. The flight time, denoted $\tau(p, q)$, for a ball thrown with initial velocity (p, q), is the time lapse from $t = 0$ to impact.

With respect to the apparent frame, the trajectory $F(t)$ of the ball will be a spiral-like arc. Let

$$M(t) = \begin{bmatrix} \cos \omega t & \sin \omega t \\ -\sin \omega t & \cos \omega t \end{bmatrix}$$

be a dynamic rotation matrix (called dynamic because it changes with time), a clockwise rotation matrix by ωt radians. Since the drum rotates at ω counterclockwise, it appears as if the ball is rotating at ω clockwise about O. Thus when the ball is at absolute position $f(t)$, the ball is at apparent position $F(t) = M(t)f(t)$. By (11),

$$F(t) = M(t) \begin{bmatrix} R \\ 0 \end{bmatrix} + t M(t) \begin{bmatrix} p \\ s \end{bmatrix} \tag{12}$$

where $s = q + \omega R$.

The first term in $F(t)$ in (12) traces out a circle of radius R, whereas the second term traces out an arithmetic polar spiral. That is, in parametric equations, the polar spiral, $r = \theta$, becomes

$$(\theta \cos \theta, \theta \sin \theta) = \theta \begin{bmatrix} \cos \theta & -\sin \theta \\ \sin \theta & \cos \theta \end{bmatrix} \begin{bmatrix} 1 \\ 0 \end{bmatrix},$$

whose line of symmetry is the y-axis. Replacing θ with $-\omega t$ results in a rescaling of the spiral as well as a reflection about the line of symmetry. Hence, as in Figure 7, $tM(t)[{}^p_s]$ is also a spiral, but has line of symmetry in the $[{}^{-s}_p]$ direction.

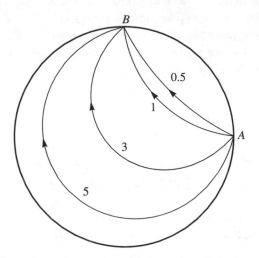

Figure 8. Apparent trajectories for various flight times τ

Before plotting apparent trajectories of balls thrown aboard Discovery, we determine the locus of velocities (p, q) for which throws go from A to B. In keeping with Figure 5, let $B = (0, R)$ in the apparent frame. A valid (p, q) velocity must satisfy

$$M(\tau)f(\tau) = \begin{bmatrix} 0 \\ R \end{bmatrix}, \tag{13}$$

where τ is the flight time $\tau(p, q)$. Since $M(-t)M(t) = I$, the identity matrix, (13) can be written as

$$f(\tau) = M(-\tau) \begin{bmatrix} 0 \\ R \end{bmatrix},$$

which gives

$$p = -\frac{R}{\tau}(1 + \sin \omega\tau) \text{ and } q = \frac{R}{\tau}(\cos \omega\tau - \omega\tau). \tag{14}$$

That is, for a given flight time τ, (14) gives the initial velocity of a ball for it to go from A to B in time τ.

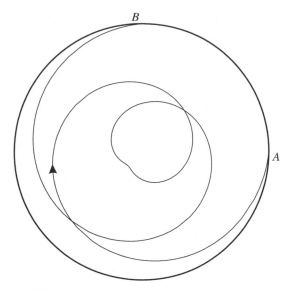

Figure 9. The apparent trajectory when $\tau = 32$

Figure 8 shows several trajectories in terms of τ. For example, the path indexed by 5 corresponds to a ball thrown from A to B in five seconds, having initial velocity $p \approx -3.5$ ft/sec and $q \approx -14.5$ ft/sec. Lest we think that these apparent trajectories are as tame as the trajectories on Asteroid B-612, the trajectory is shown in Figure 9, where $\tau = 32$ seconds. It has three loops. Since the drum makes one revolution every ten seconds, we might expect apparent trajectories to have about $\lfloor \tau/10 \rfloor$ loops, a problem we leave for the reader.

In contrast to Figure 6, the locus of velocities (p, q) allowing a ball to travel from A to B is convoluted. Figure 10 shows the graph of velocities as given by (14). The velocity $(0, -\omega R)$ is called the singular velocity. Launched with singular velocity, a ball traces out a circle of radius R in the apparent frame. The flight time in this case is infinite. In the absolute frame, the ball is motionless while the space station rotates. If a ball is thrown very fast, then its trajectory should be almost a straight line between A and B in the direction $(-1, 1)$.

If a ball is thrown so that its initial velocity is close to the singular velocity, chaos results. That is, as the flight time τ increases, minor variations in p

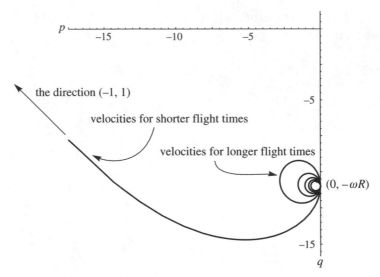

Figure 10. Locus of velocities (p, q) yielding AB flights

and q from the singular velocity result in wildly different termination points
for the apparent trajectories.

To see this better, partition the floor of the space station into four regions,
(corresponding to the four quadrants of the xy-plane), as indicated in Figure
11a. The four floor regions are painted black, dark gray, white, and light
gray. Now color each vector (p, q) with the color of the floor that the
ball hits when thrown from A at velocity (p, q). The result is shown in
Figure 11b.

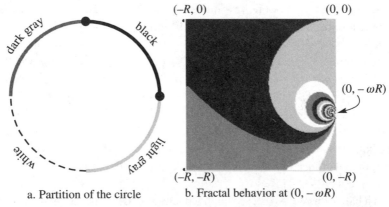

a. Partition of the circle   b. Fractal behavior at $(0, -\omega R)$

Figure 11. A fractal of velocities (p, q), with $\omega = 2\pi/10$

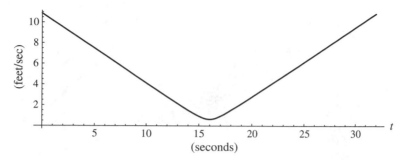

Figure 12. Speed at which the trajectory is traced, $\tau = 32$ seconds

Properties of the apparent trajectories

The trajectories of Figures 5, 8, and 9 suggest that apparent trajectories in the space station share several features in common with trajectories on the asteroid.

Property 1: The trajectories appear to be symmetric with respect to the perpendicular bisector of the chord connecting the launch point and the impact point.

Property 2: The speed of the ball decreases monotonically with the height above the ground or floor and does so symmetrically in time.

Property 3: The minimum speed occurs when the ball is furthest from the ground or floor.

On Discovery, the apparent speed of the ball is the magnitude of $F'(t)$. Figure 12 shows a graph of the speed at which the trajectory of Figure 9 is traced.

For elliptical curves such as those in Figure 5 these three properties are well known. Using a little linear algebra, we demonstrate that these same properties hold for the spiral-type curves of Figures 8 and 9. We also derive a formula for the trajectory flight time τ in terms of initial apparent velocity.

Recall that the ball is thrown from position $A = (R, 0)$ when the apparent and absolute frames agree; the ball's initial velocity is (p, q) where $p \le 0$ and q is any real number. Let $s = q + \omega R$. With respect to the absolute frame, the ball's trajectory is a chord $\overline{AB}$ of the circle for some absolute point B, giving a total flight time of τ, as in (11):

$$f(t) = \begin{bmatrix} R + pt \\ st \end{bmatrix}.$$

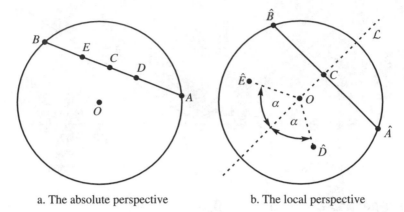

a. The absolute perspective   b. The local perspective

Figure 13. The two perspectives

Let $C = (A + B)/2$. With respect to the absolute frame, the ball moves along chord $\overline{AB}$ at constant speed. If time is translated so that the ball is at the midpoint of this line segment at $t = 0$, then, with respect to the new time variable t, the trajectory of the ball is

$$g(t) = C + t \begin{bmatrix} p \\ s \end{bmatrix},$$

where $-\tau/2 \leq t \leq \tau/2$. The trajectory Γ that the ball appears to take is given by $G(t) = M(t)g(t)$. Let $\hat{A} = G(-\tau/2)$ and $\hat{B} = G(\tau/2)$. Thus $g(0) = C = G(0)$. If $\hat{A} \neq \hat{B}$, let $\mathcal{L}$ be the perpendicular bisector of chord $\overline{\hat{A}\hat{B}}$; otherwise, let $\mathcal{L}$ be the line through O with direction $\hat{A}$. For $0 < t_0 \leq \tau/2$, let $D = g(-t_0)$, $E = g(t_0)$, $\hat{D} = G(-t_0)$, $\hat{E} = G(t_0)$, as shown in Figure 13. Since $M(-t_0)$ is a counterclockwise rotation about O by ωt_0, and, similarly, $M(t_0)$ is a clockwise rotation, then $\hat{D}$ and $\hat{E}$ are equidistant from O, and the rays $\hat{D}$ and $\hat{E}$ make the same angle with $\mathcal{L}$ (indicated by α in the figure). Thus Property 1 is verified.

Furthermore, in the absolute frame the ball is moving along the chord $\overline{AB}$ so that it is closest to O at C. Since $M(t)$ is a rotation matrix, multiplication by it leaves the magnitude of a vector unchanged, so $\|g(t)\| = \|G(t)\|$. Therefore the nearest point on Γ to O is C.

We might intuitively expect the apparent speed at which Γ is traced is the magnitude of the vector sum of (p, s) and $\omega(b, -a)$ when the ball is at absolute position (a, b), since the ball is traveling along a straight line with velocity (p, s) absolutely while the station rotates clockwise at ω radians per second. Figure 14 shows this geometry. To see whether this intuition is correct, recall that the ball's apparent position (12) is

$$F(t) = M(t) \begin{bmatrix} R + pt \\ st \end{bmatrix},$$ (15)

where $s = q + \omega R$. Its apparent velocity at t is

$$\mathbf{v}(t) = M(t) \begin{bmatrix} p \\ s \end{bmatrix} + M'(t) \begin{bmatrix} R + pt \\ st \end{bmatrix}.$$

Since $M'(t) = \omega M\big(t + \pi/(2\omega)\big) = \omega M(t) M\big(\pi/(2\omega)\big)$, we get

$$\mathbf{v}(t) = M(t) \begin{bmatrix} p \\ s \end{bmatrix} + \omega M\left(t + \frac{\pi}{2\omega}\right) \begin{bmatrix} R + pt \\ st \end{bmatrix}$$

$$= M(t) \left(\begin{bmatrix} p \\ s \end{bmatrix} + \omega \begin{bmatrix} st \\ -R - pt \end{bmatrix} \right).$$

That is, modulo a rotation, $\mathbf{v}$ has the structure we imagined.

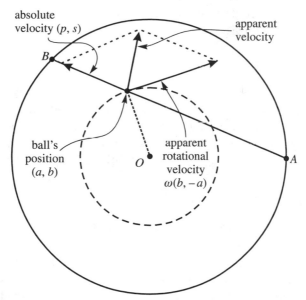

Figure 14. An intuitive decomposition of the apparent velocity of the ball

Let's now calculate the time at which the ball is nearest to O, which is also the time at which the ball is at C. At this time, the apparent velocity of the ball is perpendicular to its position vector. Since the ball's apparent position at time t is $F(t)$ and its apparent velocity is $F'(t)$, we solve for the time at which $F(t) \cdot F'(t) = 0$. By (15) and (1) this is

$$\begin{bmatrix} R + pt, & st \end{bmatrix} M(t)^\mathsf{T} M(t) \begin{bmatrix} p + \omega st \\ s - \omega R - \omega pt \end{bmatrix} = 0.$$

Since $M(t)^\mathsf{T} M(t)$ is the identity matrix, simplification gives $pR + p^2 t + q^2 t = 0$, which means that the time it takes the ball to reach the midpoint of the chord $\overline{AB}$ is

$$\frac{\tau}{2} = -\frac{pR}{p^2 + s^2}. \tag{17}$$

Thus, the flight time of the throw from A to B is

$$\tau = -\frac{2pR}{p^2 + (q + \omega R)^2}, \tag{18}$$

and the ball hits the floor at relative position $F(\tau)$. We suspect that $\mathbf{v}(t)$ has minimum magnitude at time $\tau/2$. To see this, let $w(t) = ||\mathbf{v}||^2$. By (1), since multiplying by $M(t)$ leaves the magnitude of a vector unchanged,

$$w(t) = (p + \omega st)^2 + (s - \omega R - \omega pt)^2.$$

This quadratic has a minimum at $t = -pR/(p^2 + s^2)$, which by (17) means that the ball's speed is least at this critical time, $\tau/2$. Furthermore, the ball's speed decreases monotonically from time 0 to $\tau/2$ and increases monotonically from $\tau/2$ to τ and does so symmetrically with respect to time $\tau/2$. And so Properties 2 and 3 are verified.

Throwing in three dimensions

Now imagine throwing a ball laterally in Discovery's rotating drum. Since the distance between home plate and pitcher's mound is 60.5 ft in the game of baseball, imagine that the drum is a cylinder 35 feet in diameter and over 60.5 feet long so that we can simulate moon-type conditions as on a lunar baseball diamond. Let the axis of rotation be the z-axis. When a ball is thrown with initial velocity (p, q, z), the third coordinate of the ball's velocity remains constant both in the apparent and absolute frames. That is, if the ball doesn't hit the side of the cylinder first, it reaches the catcher in $60.5/z$ seconds. We adapt the trajectory of Figure 9 to three dimensions, and throw a ball from a pitcher's mound at $(R, 0, 0)$ to home plate at $(0, R, 60.5)$. We let $p = -(R/32)(1 + \sin(32\omega))$, $q = (R/32)(\cos(32\omega) - 32\omega)$ by (14), and $z = 60.5/32$. Figure 15 displays the trajectory of the pitch from looking down the central axis of the cylindrical space station. A pitcher can throw quite a curve ball on board Discovery! (In Figure 15, the radius of the cylinder is taken to be 20 feet so as to allow both the pitcher and the catcher to throw and catch the ball 2.5 feet above the floor.)

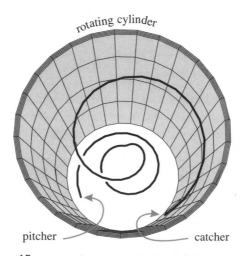

Figure 15. A sweeping curve ball, from pitcher to catcher

For a different development of these results using mechanics see Fowles [**33**, pp. 171–176], but be prepared for solving many differential equations! The little bit of linear algebra we've used in the chapter has done surprisingly well against the competition.

It may be that in years to come the Olympics will involve sports in space. Perhaps field games will be played within rotating cylinders. In this environment, space athletes must develop a sense of the speed and direction with which to throw or kick balls to teammates. On earth that is quite intuitive—just throw the ball towards the teammate. Within a rotating arena that sense is less intuitive. Somehow ballplayers must develop the knack of implicitly solving the equation $M(\tau)f(\tau) = B$ for any flight time τ and any position B, so that balls reach their targets.

Exercises

1. Imagine an asteroid of radius R whose gravitational acceleration at its surface is 32 ft/sec^2. Find the relation between R and the escape velocity from the asteroid's surface. How large does R need to be in order to sustain an atmosphere where winds reach up to 155 miles per hour, the speed of category 5 hurricanes?

2. Suppose that Asteroid B-612 rotates once a minute. Find the trajectory of a ball thrown from the equator to the north pole, and vice versa. Are the flight times the same?

3. Let $A = (R, 0)$ and $B = (0, R)$. On Discovery, which is easier, throwing a ball from A to B or from B to A?

4. Using a mathematical software system, make a movie of the trajectories of a ball thrown from A to B inside Discovery as the flight time τ increases. When does a loop appear in a trajectory?

5. Simplify Figure 11 by splitting the circle into two arcs by cutting along the x-axis. For any initial velocity vector (p, q) of a ball thrown from A, paint point (p, q) either black or white depending on whether the ball first impacts above or below the x-axis, respectively.

6. Given two spirals,

$$S_1(t) = t M(t) \begin{bmatrix} p_1 \\ q_1 \end{bmatrix}$$

and

$$S_2(t) = t M(t) \begin{bmatrix} p_2 \\ q_2 \end{bmatrix},$$

determine the axis of symmetry of $\lambda S_1(t) + \mu S_2(t)$.

7. Given the circle $M(t)[\begin{smallmatrix}1\\0\end{smallmatrix}]$ and the spiral $t M(t)[\begin{smallmatrix}p\\q\end{smallmatrix}]$, determine the axis of symmetry of their sum.

8. Imagine throwing a ball with velocity (p, q) from position $(R, 0)$ in a nonrotating space station, a disk of radius R centered at the origin. Find the ball's flight time without first finding the impact point.

9. Solve the problem of determining a ball's trajectory on Discovery, not neglecting air resistance.

10. Explain how trajectories on Discovery change as R and ω change.

preamble xiii

Adopt the character of the twisting octopus, which takes on the appearance of the nearby rock. Now follow in this direction, now turn a different hue.

Theognis (circa 545, BC), Elegies

"Have you considered the rotating beacon problem?"

The questioner was from a sister school over the mountains in North Carolina. I had just finished giving a talk on the falling ladder problem of Chapter V at a session of the Southeast Section of the Mathematical Association of America.

The beacon problem is familiar. It's a related rates exercise from first year calculus whose standard solution yields the implication that some things travel faster than light, as happens also in the ladder problem.

I confessed my ignorance, but the question continued to bother me. A few months later, having jotted down some ideas and formulas, I wrote to Irl Bivens, who had asked the question, inviting him to a joint attempt at resolving the paradox of the problem.

Thus began a two year correspondence, ultimately filling five bulging notebook binders, mostly by e-mail, in which we generated curves, proposed conjectures, critiqued each other's arguments, and felt our way through the jungle-like mystery of a neglected mathematical construct: the envelope of a family of curves.

While working out the details of the beacon problem, I learned afresh that two are more powerful than one. To have a friend who is as keen as you are about an idea, to have a knowledgeable, attentive, independent mind on which to test ideas: with such teamwork, problems that are perhaps outside one's ability to resolve now become sportingly good fun.

This next chapter is a recasting of our joint paper [11] on the beacon problem. Some of the phrases used are Irl's. However rather than listing him as a joint author of this chapter, I've taken the liberty of naming a transformation and a technique after him: the Bivens transform is a two-point normalization of an ellipse, and the Bivens parametrization is a simple way to parametrize the envelope of some families of curves. One cannot in good taste name an idea after himself. Another must do it for him. It is my pleasure to do that for my friend.

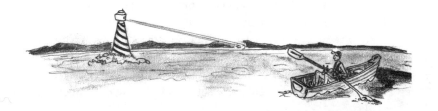

CHAPTER XIII

The Rotating Beacon

This chapter is about a phenomenon that occurs when viewing light from a rotating beacon striking a surface, which may seem to be out of place in a book devoted to falling. Yet a light beam can be perceived as particles, a stream of photons that are emitted from a source and fly (or fall) through space, until encountering matter, there to be absorbed or reflected. In that sense, this chapter is at home in this book. The phenomenon we discuss starts with a problem almost everyone who has taken a calculus course has solved and ends with the analysis of an exotic curve.

Here's the phenomenon, which we state as a question:

Imagine a powerful beacon positioned on an island, its beam illuminating the convex shoreline of a lake as it rotates counterclockwise with constant angular velocity ω. Is there any place on the lake where a viewer could sit in a rowboat so that the illuminated spot on the beach always appears to move about the shoreline in a counterclockwise direction?

The reader may be inclined to think that nothing to the contrary could occur, that any place on the lake is equally satisfactory. However if ω is large enough and if we take into account the finite speed of light then unexpected behavior results.

For such a light show, we will call ordinary those vantage points in the lake where a viewer sees the beacon's light show as ever counterclockwise. We show that for moderate rotation rates the set of ordinary points is a convex subset of the region bounded by the shoreline. In all cases save one, the set of ordinary points will be empty at high rotation rates, the exception being an elliptical shoreline with the beacon at one focus. In this case the set of ordinary points always contains the second focus of the ellipse. As the period of the beacon decreases, the set of ordinary points shrinks to the second focus at a rate that (to first order) is proportional to the period of the beacon.

Figure 1. The lighthouse at Alexandria, the seventh wonder of the ancient world

We show that the shape of the set of ordinary points converges to the body of a fish-shaped curve called the antiorthotomic of an ellipse. Doing this, we present a way of generating envelopes of a family of lines that differs from the classical approach of Chapter X, and may prove useful as a general mathematical technique.

The original problem

Fancifully pictured in Figure 1, the lighthouse at Alexandria, one of the fabled seven wonders of the ancient world, was built on the island of Pharos outside the harbor of Alexandria. Constructed in about 290 BC with a statue of Poseidon gracing its crest, it stood 384 feet high, the height of a forty story modern day skyscraper, and lasted 1500 years. Reflecting the sun's rays by day and a fire by night using mirrors, the lighthouse could be seen at least 35 miles out at sea. Its beacon could be focused at will—according to legend, focused sunshine from its mirrors was enough to set enemy ships ablaze—and could be swept along a region of the sea or along the coastline. Perhaps it was from just such a practice that people first posed the rotating beacon problem, a problem that now is found in the related rates section of almost every beginning calculus text:

> A beacon rotating counterclockwise at ω radians per second is one mile from shore. How quickly does the illuminated spot on the shoreline move along the beach?

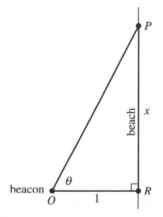

Figure 2. The beacon O and the illuminated spot P, (infinite speed of light model)

In this section, we review the standard solution to this problem, and point out some shortcomings in its underlying model. Then we show how a better model overcomes them, at the price of generating some peculiar behavior of its own.

Assume the beacon is located at point O in Figure 2. Assuming that the beach is a straight (vertical) line, let x be the signed distance between the illuminated spot P and the point R on the beach closest to the beacon, where the positive x direction is up. Let θ denote the angle between the direction of the beacon and ray OR at time t. Texts assume that the speed of light is infinite, so at any time there will be a unique illuminated spot on the beach if and only if the beacon is pointing towards the shoreline at that moment. We then have $x = \tan\theta$ and by the chain rule

$$\frac{dx}{dt} = \frac{dx}{d\theta}\frac{d\theta}{dt} = \omega\sec^2\theta = \omega(1 + x^2). \tag{1}$$

So, the position x of the illuminated spot is a strictly increasing function of t and its velocity $\frac{dx}{dt}$ becomes infinite as $|x|$ approaches infinity. Likewise, the velocity of the illuminated spot becomes infinite as ω approaches infinity.

A problem with this solution is that the speed of light, c, is finite. Of course, for all practical terrestrial problems this does not matter. Nonetheless, let us model the problem using the assumption of finite light speed.

The time for light to travel from O to P is $\sqrt{1 + x^2}/c$. Thus at the moment when spot P is illuminated, the beacon no longer points towards P; it has rotated $\omega\sqrt{1 + x^2}/c$ counterclockwise away from that direction. Hence, the expression for the angle $\theta(x)$ of the beacon when P is illuminated x miles along the shore line is

$$\theta(x) = \tan^{-1} x + \frac{\omega\sqrt{1 + x^2}}{c}. \tag{2}$$

Using the chain rule again, namely, $\frac{dx}{dt} = \frac{d\theta}{dt}/\frac{d\theta}{dx} = \omega/\frac{d\theta}{dx}$, and differentiating (2) gives

$$\frac{dx}{dt} = \frac{c\omega(1 + x^2)}{c + \omega x\sqrt{1 + x^2}}, \tag{3}$$

as was first noted by Althoen [3].

The denominator of (3) vanishes at $x_0 = -\sqrt{\sqrt{1/4 + (c/\omega)^2} - 1/2}$ so that the velocity of the spot is undefined at x_0. For example, if the beacon is rotating at one revolution per minute then $x_0 \approx -1333$ miles. Geometrically, x_0 is the point at which the spiral wavefront of light from the beacon initially splashes onto the beach, as shown in Figure 3.

The velocity of the spot is negative for $x < x_0$ and is positive for $x > x_0$. Thus, contrary to the infinite light speed solution, the spot on the shoreline is actually moving in the negative x direction for $x < x_0$. A similar phenomenon sometimes appears in old war movies: a machine gunner swings his machine gun in an arc from a position parallel to a wall toward the wall while firing the gun, and the bullets strike the wall in the reverse direction, with the track of bullet holes traveling away from the gun. The reason this happens is that the bullets fired first are virtually parallel to the wall and take a relatively long time to reach it. Bullets fired later have a shorter distance to travel and strike

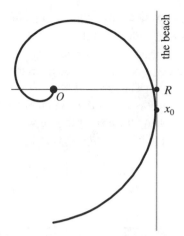

Figure 3. A wave front rolls onto the beach

the wall before the earlier bullets. A safer demonstration of this phenomenon can be carried out with a rotating lawn sprinkler.

If we allow c to approach infinity in (3) then we recover the solution of (1). As $|x|$ approaches infinity, the speed of the spot does not become infinite but instead approaches c. On the other hand, near the point x_0 the speed of the spot does become arbitrarily large. But this poses no physical impossibility, as it is the position of illumination that we are tracking; no physical particle is moving along in tandem with that position of illumination. For a further discussion of this idea, see Rothman [**67**]. As ω approaches infinity, the initial contact point x_0 approaches 0 and the velocity of the spot at $x \neq 0$ approaches $c\sqrt{1 + x^2}/x$. Although this limiting velocity is greater than c, it is still finite. If we assume the beacon has been rotating for all time, then at any given moment there will be infinitely many spots moving along the shoreline in both directions. These conclusions are in dramatic opposition to those of the infinite light speed model.

The beacon on an island in a convex lake

Now imagine that the beach is a smooth convex closed curve. We choose a coordinate system so that the beacon is at the origin and the beach is described by a polar graph $r = f(\theta) > 0$. To simplify the formulas we choose units such that the speed of light c is equal to 1, so it takes $f(\theta)$ time units for light to travel $f(\theta)$ distance units. Assume that at time $t = 0$ the beacon is pointing in the direction $\theta = 0$. The time at which the illuminated spot is at

$r = f(\theta)$ is $t = t(\theta) = \theta/\omega + f(\theta)$. That is, it takes θ/ω time units for the beacon to point in the θ direction, and then $f(\theta)$ time units for light to reach the beach. By the familiar arclength formula, $\frac{ds}{d\theta} = \sqrt{(f(\theta))^2 + (f'(\theta))^2}$, where $s = s(\theta)$ is the distance measured counterclockwise along the beach from $r = f(0)$ to $r = f(\theta)$. Then $\frac{dt}{d\theta} = 1/\omega + f'(\theta)$ so that on intervals of θ for which $f'(\theta) \neq -1/\omega$, θ is a smooth function of t with implicit derivative $\frac{d\theta}{dt} = \omega/(1 + \omega f'(\theta))$. Using the chain rule, we see that the velocity $v(\theta)$ of the spot on the beach at point $r = f(\theta)$ is given by

$$v(\theta) = \frac{ds}{dt} = \frac{ds}{d\theta}\frac{d\theta}{dt} = \frac{\omega\sqrt{[f(\theta)]^2 + [f'(\theta)]^2}}{1 + \omega f'(\theta)}. \qquad (4)$$

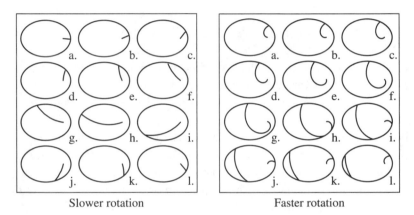

Slower rotation   Faster rotation

Figure 4. Once around

Since the sign of $v(\theta)$ is the same as that of $\frac{d\theta}{dt}$ we see that the spot is moving counterclockwise around the beach when $v(\theta) > 0$ and clockwise when $v(\theta) < 0$. As with a straight beach, there may be more than one illuminated spot on the beach at a time. To understand this phenomenon better, consider Figure 4. Each side of the figure contains a succession of twelve snapshots of the beacon's wavefront in one complete revolution. The time lapse between successive snapshots is the same. The snapshots on the left correspond to a slow rotation rate for the beacon. Only one spot on the beach is illuminated at any time, and the spot is always moving counterclockwise. The snapshots on the right correspond to a fast rotation rate for the beacon and illustrate that as the rotation rate increases, a qualitative change in the behavior of the illuminated spot takes place. Until (g) there is a single illuminated spot on the beach. Then between snapshot (g) and snapshot (h) a wavefront splashes onto the beach and three points on the beach are illuminated, two moving counterclockwise and the other clockwise. Shortly after snapshot (l) two of the illuminated spots collide (at infinite speed) and disappear at one end of the beach while the third continues moving counterclockwise at the other end.

As the rotation rate increases, more and more points along the beach will be illuminated concurrently, so that as $\omega \to \infty$ the entire beach is ablaze with light continually moving in both directions. Even more complications can arise. For example, if $f'(\theta) = -1/\omega$ on some interval, then light from the beacon arrives at all the corresponding points $r = f(\theta)$ simultaneously. Geometrically this means that a portion of the beach coincides with the wavefront of the beacon, so it becomes difficult to define the position of the spot on the beach. Such a light show is in stark contrast to what one might naïvely expect, a single spot of light moving counterclockwise around the beach.

In assigning space and time coordinates to physical events it is important to distinguish between observing an event and seeing an event. An observer is assumed to be omniscient and omnipresent, knowing at every instant what is happening anywhere within the observer's frame of reference. But for someone to see an event, light must travel from the location of the event to the eyes of the viewer. Because of this optical delay, what is happening and what appears to be happening can be very different. When we use the words *see*, *view*, and *appear* we mean seeing, not observing.

The snapshots of Figure 4 correspond to observing a spot of light on the beach. For the rest of this chapter we abandon such a perspective unless stated otherwise, and adopt the perspective of seeing a spot of light on the beach.

We define a vantage point within the region bounded by the beach to be ordinary for a particular rotation rate if the view of the beach from that point always shows a single spot of light moving with finite speed counterclockwise around the beach. A vantage point that is not ordinary we will call extraordinary.

Alternate ways of being ordinary

In this section we look at different ways for a point to be ordinary.

We assume that the beacon is at the origin O and that the beach is given by the polar curve $r = f(\theta) > 0$ with period 2π and has a regular parametrization, $g(\theta) = f(\theta)(\cos\theta, \sin\theta)$, so $\|g'(\theta)\| > 0$ for all θ. Let $A = (a, b)$ be the position of a viewer in a boat on the lake bounded by f. We say that A sees a spot at $g(\theta)$ if the viewer at A sees the spot as illuminated. Thus our first perspective is the definition: A is an ordinary point in the lake if

i. definition of ordinary:

> From A, just one spot is seen to move, ever counterclockwise.   (5)

Let $h_A(\theta)$ be the distance from A to $g(\theta)$:

$$h_A(\theta) = \|A - g(\theta)\| = \sqrt{(a - f(\theta)\cos\theta)^2 + (b - f(\theta)\sin\theta)^2}.$$

Then,

$$h'_A(\theta) = \frac{dh_A(\theta)}{d\theta} = -\frac{g'(\theta) \cdot (A - g(\theta))}{h_A(\theta)}. \tag{6}$$

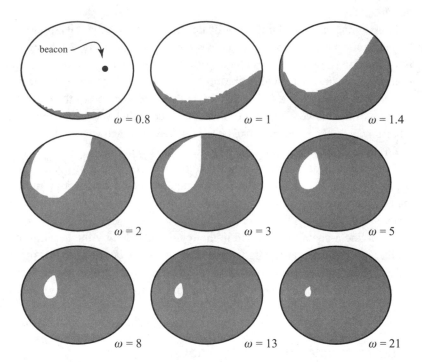

beacon

$\omega = 0.8$ $\omega = 1$ $\omega = 1.4$

$\omega = 2$ $\omega = 3$ $\omega = 5$

$\omega = 8$ $\omega = 13$ $\omega = 21$

Figure 5. The vanishing ellipse, primitive point version

The time $t_A(\theta)$ for A to see the spot at $g(\theta)$ is

$$t_A(\theta) = \frac{\theta}{\omega} + f(\theta) + h_A(\theta). \tag{7}$$

That is, it takes θ/ω for the beacon to point towards θ, $f(\theta)$ for the light from the beacon to reach the beach at $g(\theta)$, and $h_A(\theta)$ for the light reflected from the beach to reach the viewer. Remember that units have been chosen so that the speed of light is $c = 1$. Thus another way to say that A is an ordinary point with respect to the rotation rate ω is

ii. time goes forward: $\boxed{t'_A(\theta) > 0 \text{ for all } \theta.}$ (8)

Figure 6. The vanishing Cheshire Cat [**18**, p. 67]

That is, A being ordinary means that A sees the illuminated spot as never stopping but always moving counterclockwise about the beach, which is the same as saying that as θ increases so does $t_A(\theta)$. At an ordinary point, $t'_A(\theta)$ is never zero. To see this, let $s(t)$ be the arclength along the beach to where the illuminated spot appears to be. Since A is ordinary and $g(\theta)$ is regular, ds/dt is finite and

$$\frac{ds}{dt} t'_A(\theta) = \frac{ds}{dt}\frac{dt}{d\theta} = \frac{ds}{d\theta} = \sqrt{(f(\theta))^2 + (f'(\theta))^2} = \|g'(\theta)\| \neq 0,$$

so $t'_A(\theta) \neq 0$ for any θ.

By (6) and (7),

$$t'_\mathbf{A}(\theta) = \frac{dt}{d\theta} = \frac{1}{\omega} + f'(\theta) - \frac{\mathbf{g}'(\theta) \cdot (\mathbf{A} - \mathbf{g}(\theta))}{h_\mathbf{A}(\theta)}. \quad (9)$$

We can use (8) and (9) to depict the set of ordinary points. Figure 5 consists of nine snapshots of the elliptical lake, $r = 1/(2 + \cos\theta)$, as the rotation

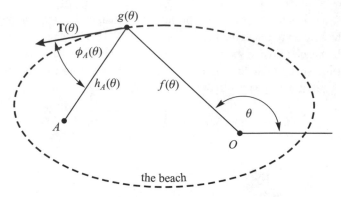

Figure 7. Perspective from a boat at A

rate ω of the beacon increases, where the beacon is situated at the right-hand focus. The shaded points in the lake represent extraordinary points. Thus, when ω is slow, as $\omega = 0.8$, almost every point in the lake is ordinary. However, as ω increases the set of regular points seems to collapse to a region clustered about the left-hand focus of the elliptical lake.

Similarly, the set of ordinary points for any lake and any beacon on the lake will vanish as the rotation rate increases, reminiscent of the Cheshire Cat of the Alice stories, wherein its extremities disappear until only its smile remains, whereupon that too vanishes (see Figure 6).

> [The Cat] vanished quite slowly, beginning with the end of the tail, and ending with the grin, which remained some time after the rest of it had gone [**18**, p. 67].

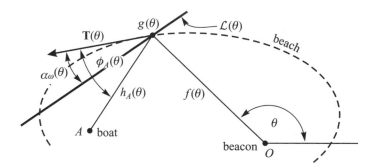

Figure 8. The line of separation $\mathcal{L}(\theta)$

We refine (8) by focusing on the angle between the tangent to the beach and the boat from $g(\theta)$. Let $\mathbf{T}(\theta)$ be the unit tangent direction to the beach at θ, so

$\mathbf{T}(\theta) = g'(\theta)/||g'(\theta)||$. Let $\phi_A(\theta)$ be the angle between $\mathbf{T}(\theta)$ and $A - g(\theta)$. Figure 7 shows the relation between these angles. Since we assume that A is on the lake, not on the beach, $0 < \phi_A(\theta) < \pi$. By (3) of the appendix and (6),

$$\cos \phi_A(\theta) = \frac{g'(\theta)}{||g'(\theta)||} \cdot \frac{A - g(\theta)}{||A - g(\theta)||} = -\frac{h'_A(\theta)}{||g'(\theta)||}. \tag{10}$$

Thus $h'_A(\theta) = -||g'(\theta)|| \cos \phi_A(\theta)$, so (9) can be written as

$$t'_A(\theta) = 1/\omega + f'(\theta) - ||g'(\theta)|| \cos \phi_A(\theta).$$

Thus (8) is equivalent to

***iii.* angle to viewer:** $\boxed{\cos \phi_A(\theta) < \dfrac{1/\omega + f'(\theta)}{||g'(\theta)||} \text{ for all } \theta.}$ (11)

To translate the algebraic expression (11) into an equivalent geometric one, define the angle $\alpha_\omega(\theta)$, where $0 \leq \alpha_\omega(\theta) < \pi$, so that

$$\cos \alpha_\omega(\theta) = \begin{cases} \dfrac{1/\omega + f'(\theta)}{||g'(\theta)||}, & \text{if } 1/\omega + f'(\theta) < ||g'(\theta)||, \\ 1, & \text{otherwise.} \end{cases} \tag{12}$$

(Since $\omega > 0$ and $||g'(\theta)|| = \sqrt{f^2(\theta) + (f'(\theta))^2}$, then $(1/\omega + f'(\theta))/||g'(\theta)|| > -1$ for all θ, so $\alpha_\omega(\theta) < \pi$.)

Thus (11) becomes $\cos \phi_A(\theta) < \cos \alpha_\omega(\theta)$. Because the inverse cosine function is decreasing, $\phi_A(\theta) > \alpha_\omega(\theta)$. Now let $\mathcal{L}(\theta)$ be the line through $g(\theta)$ containing the direction obtained by rotating $\mathbf{T}(\theta)$ counterclockwise by $\alpha_\omega(\theta)$. We say that a point B is strictly on one side of a line if B is not on the line. Thus the line $\mathcal{L}(\theta)$ separates the point A and the ray $\{g(\theta) + u\mathbf{T}(\theta) | u \geq 0\}$ for any given θ; that is, A is strictly on one side of $\mathcal{L}(\theta)$ and $g(\theta) + \mathbf{T}(\theta)$ is on the other side. Figure 8 shows the arrangement of $\mathcal{L}(\theta)$, A, and $g(\theta) + \mathbf{T}(\theta)$. Therefore (11) is equivalent to

***iv.* separating line:** $\boxed{\mathcal{L}(\theta) \text{ separates } A \text{ and } \{g(\theta) + u\,\mathbf{T}(\theta) | u \geq 0\} \text{ for all } \theta.}$ (13)

We can use (13) to depict the set of ordinary points. Figure 9 shows a series of nine snapshots of the elliptical lake, $r = 1/(2 + \cos \theta)$, as the rotation rate ω of the beacon increases, where the beacon is situated at the right-hand focus. The region within the lake not penetrated by any of the separating

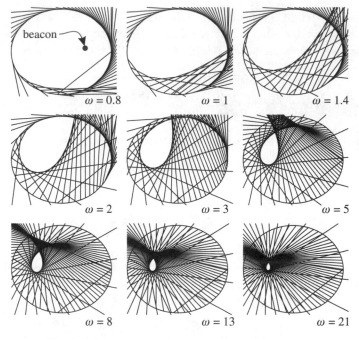

Figure 9. The vanishing ellipse, separating lines version

lines is the set of ordinary points. As compared with Figure 5, this figure is a clearer and quicker way of displaying the set of ordinary points within the lake.

Another way to recast the notion of ordinariness is by (4) and (11),

$$
\boxed{\textbf{\textit{v.} Doppler version:} \quad \text{For each } \theta, \begin{cases} v(\theta)\cos\phi_A(\theta) < 1, \text{ if } v(\theta) > 0 \\ v(\theta)\cos\phi_A(\theta) > 1, \text{ if } v(\theta) < 0 \end{cases}}
$$

$$\tag{14}$$

We give a thought experiment explanation of this Doppler perspective. By (4), $v(\theta)$ is never 0. We ignore the case when $v(\theta)$ is undefined, that is, when $f'(\theta) = -1/\omega$. To see why we call (14) the Doppler perspective, recall that $v(\theta)$ is the observed speed of the illuminated spot along the beach. Since the viewer sees the spot from an angle at A, its apparent speed is $v(\theta)\cos\phi_A(\theta)$. Imagine a clock moving along with the illuminated spot. If $v(\theta) > 0$, then the clock appears to approach the viewer at speed $v(\theta)\cos_A(\theta)$. If this apparent speed is less than 1, then light leaving the clock one moment must travel a greater distance to reach the viewer than light leaving the clock the next moment, so time, according to the viewer watching the clock, appears to speed up. If the apparent speed exceeds 1—if the clock could approach the viewer at a speed greater than the speed of light—then the clock would appear to be running backwards, because the light from a later moment would be seen before the light from an earlier moment. The clock would get ahead of light reflected from itself in the past! Similarly, if $v(\theta) < 0$ and $v(\theta) \cos_A(\theta) > 1$, then $v(\theta) < -1$ and the illuminated clock runs clockwise along the beach. But the clock appears to move towards the viewer at a speed greater than 1. Hence the illuminated spot appears to travel counterclockwise while the clock itself appears to run backward!

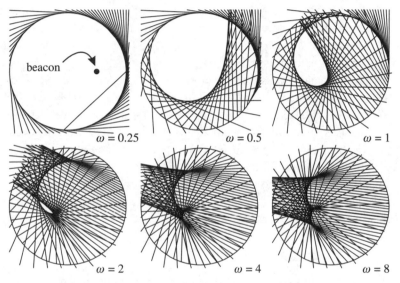

Figure 10. A beacon at $(0, 0)$ with shoreline $(x + 1)^2 + y^2 = 4$

Properties of the ordinary region

For a given rotation rate, let the ordinary region of a lake be the set of all its ordinary points. Figures 5 and 9 suggest some general characteristics of the ordinary region.

First of all, the ordinary region for a given rotation rate is a convex set. This follows immediately from (13), since for each θ the ordinary region always lies on one side of the separating line $\mathcal{L}(\theta)$, and the intersection of any number of half planes is a convex set.

Second, when the lake is an ellipse and the beacon is at a focus, then the ordinary region collapses to the other focus as the rotation rate of the beacon increases. This follows almost immediately from (8). Let F_1 and F_2 be the two foci of an ellipse, with the beacon positioned at F_2, the origin. Position the boat at F_1. The defining property of an ellipse is that the sum of the distances from F_2 to any point B on the ellipse and from B to F_1 is a constant. That is, $f(\theta) + h_{F_1}(\theta)$ is a constant. By (7), $t'_{F_1}(\theta) = 1/\omega > 0$. By (8), for all ω,

$$F_1 \text{ is always an ordinary point.} \tag{15}$$

Now let A be any point in the lake other than F_2, and let $q(\theta) = f(\theta) + h_A(\theta)$. Since A is not the second focus, $q(\theta)$ is a nonconstant periodic function, and so will sometimes be decreasing. Let θ_0 be an angle for which $q'(\theta_0)$ is negative. Choose ω such that $1/\omega < -q'(\theta_0)$. Then $t'_A(\theta_0) < 0$, which means that A is extraordinary.

Figure 10 contains six snapshots of the ordinary region for a lake whose beach is a circle centered at $(-1, 0)$ with radius 2, and whose beacon is at the origin. For this lake, $f(\theta) = -\cos\theta + \sqrt{3 + \cos^2\theta}$. As can be seen, as ω increases the ordinary region collapses to the empty set. This occurs whenever the beacon is positioned at any point in a non-elliptical lake, a general result we leave for the reader to verify.

Finally, observe that no ordinary point can be on a separating line. This means that the boundary of the ordinary region is made up of extraordinary points. Finding it is our next objective.

The Bivens parametrization

In this section we consider a way to find the envelope to a family of separating lines, such as depicted in Figures 9 and 10. Let $\mathbf{T}(\theta)$ be the unit tangent vector to $g(\theta)$, and let $\mathbf{N}(\theta)$, being $\mathbf{T}(\theta)$ rotated counterclockwise by $\pi/2$, be the unit normal to the curve. Let $\alpha(\theta)$ be a smooth function, which we will interpret as giving the angular counterclockwise rotation of a separating line through $g(\theta)$ away from $\mathbf{T}(\theta)$. To decompose the direction of a separating line into the two natural directions $\mathbf{T}(\theta)$ and $\mathbf{N}(\theta)$, we let

$$\mathbf{W}_1(\theta) = \cos\alpha(\theta)\,\mathbf{T}(\theta) + \sin\alpha(\theta)\,\mathbf{N}(\theta). \tag{16}$$

See Figure 11 for the geometrical relationships of these vectors. In calculus texts, the $\mathbf{T}(\theta)$ and $\mathbf{N}(\theta)$ axes are often referred to as the Serret Frenet coordinate system. Let $\mathcal{F}$ be the family of lines $\{g(\theta) + u\,\mathbf{W}_1(\theta) \,|\, u \in \mathcal{R}\}_{\theta \in \mathcal{R}}$.

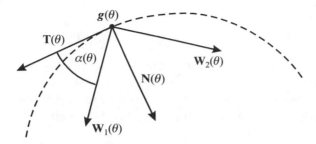

Figure 11. The Serret Frenet coordinate system for a unit-speed curve

The normal direction to $\mathbf{W}_1(\theta)$, a rotation of $\mathbf{W}_1(\theta)$ by $\pi/2$ radians counter-clockwise, is given by

$$\mathbf{W}_2(\theta) = -\sin\alpha(\theta)\,\mathbf{T}(\theta) + \cos\alpha(\theta)\,\mathbf{N}(\theta). \tag{17}$$

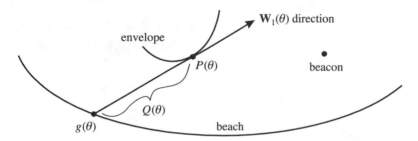

Figure 12. The Bivens parametrization of an envelope

As shown in Exercise 1 or in Bruce [15, pp. 26–8], the curvature $\kappa(\theta)$ of the curve $g(\theta) = (x(\theta), y(\theta))$ is

$$\kappa(\theta) = \frac{x'(\theta)y''(\theta) - x''(\theta)y'(\theta)}{\|g'(\theta)\|^3}. \tag{18}$$

Furthermore,

$$\mathbf{T}'(\theta) = \kappa(\theta)\|g'(\theta)\|\,\mathbf{N}(\theta) \quad \text{and} \quad \mathbf{N}'(\theta) = -\kappa(\theta)\|g'(\theta)\|\,\mathbf{T}(\theta). \tag{19}$$

Therefore, by (16) and (19),

$$\begin{aligned}
\mathbf{W}_1'(\theta) &= -\sin\alpha\,\alpha'\,\mathbf{T} + \cos\alpha\,\kappa\,\|g'\|\,\mathbf{N} + \cos\alpha\,\alpha'\,\mathbf{N} - \sin\alpha\,\kappa\,\|g'\|\,\mathbf{T} \\
&= \big(\kappa(\theta)\|g'(\theta)\| + \alpha'(\theta)\big)\big(-\sin\alpha(\theta)\,\mathbf{T}(\theta) + \cos\alpha(\theta)\,\mathbf{N}(\theta)\big) \\
&= \big(\kappa(\theta)\|g'(\theta)\| + \alpha'(\theta)\big)\,\mathbf{W}_2(\theta). \tag{20}
\end{aligned}$$

By (16) and (17),

$$\mathbf{T}(\theta) = \cos\alpha(\theta)\,\mathbf{W}_1(\theta) - \sin\alpha(\theta)\,\mathbf{W}_2(\theta). \tag{21}$$

The envelope for $\mathcal{F}$ is a parametrized curve $P(\theta)$ where $\{g(\theta_0) + u\,\mathbf{W}_1(\theta_0)\,|\,u \in \mathcal{R}\}$ is a tangent line to the curve at the point $P(\theta_0)$ for every θ_0. Thus $P(\theta) = g(\theta) + Q(\theta)\,\mathbf{W}_1(\theta)$ for some function $Q(\theta)$ with the property that the vector $P'(\theta)$ points in the $\mathbf{W}_1(\theta)$ direction. Figure 12 shows these relationships.

To find $Q(\theta)$, from $g'(\theta) = ||g'(\theta)||\,\mathbf{T}(\theta)$, (20), and (21),

$$P'(\theta) = g(\theta) + Q'(\theta)\,\mathbf{W}_1(\theta) + \big(\kappa(\theta)||g'(\theta)|| + \alpha'(\theta)\big)Q(\theta)\,\mathbf{W}_2(\theta)$$
$$= (Q' + \cos\alpha\,||g'||)\mathbf{W}_1 + \big((\kappa||g'|| + \alpha')Q - \sin\alpha\,||g'||\big)\,\mathbf{W}_2.$$

Hence, for $P'(\theta)$ to point in the $\mathbf{W}_1(\theta)$ direction,

$$\big(\kappa(\theta)||g'(\theta)|| + \alpha'(\theta)\big)Q(\theta) - \sin\alpha(\theta)||g'(\theta)|| = 0. \tag{22}$$

If we ignore the singular case when $\kappa(\theta)||g'(\theta)|| + \alpha'(\theta) = 0$, then (22) gives $Q(\theta) = ||g'(\theta)||\sin\alpha(\theta)/(\kappa(\theta)||g'(\theta)|| + \alpha'(\theta))$. Thus the Bivens parametrization of the envelope to $\mathcal{F}$ is

$$P(\theta) = g(\theta) + Q(\theta)\mathbf{W}_1(\theta), \quad \text{where} \quad Q(\theta) = \frac{||g'(\theta)||\sin\alpha(\theta)}{\kappa(\theta)||g'(\theta)|| + \alpha'(\theta)}. \tag{23}$$

To find the envelopes corresponding to Figure 9, let $\cos\alpha(\theta) = \cos\alpha_\omega(\theta)$ of (12). Thus $\sin\alpha(\theta)$ and $\alpha'(\theta)$ are defined. Figure 13 gives the resulting envelopes corresponding to Figures 5 and 9. As can be seen the envelope appears to be a fish-shaped curve. The points within the body are ordinary points, but those in the tail are not.

In Exercise 9, we give another application of the Bivens parametrization and show a different way of generating the epicycloids of Chapter X as envelopes of a family of lines.

Our next goal is to see that the shape of the envelope of Figure 13 converges to that of a classic curve, but first, we include a short interlude about ellipses.

An elliptical interlude

In order to see the relationship between the envelope in the rotating beacon problem and a well-known curve, we will consider ways of generating new curves from old ones. We look at three such transformations in this section:

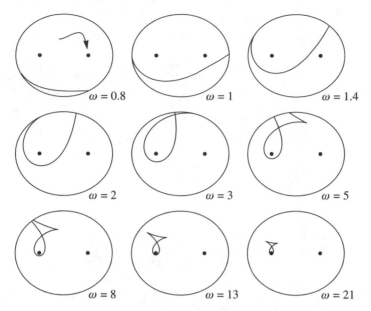

Figure 13. The vanishing ellipse, envelope version

a two-point normalization of a curve, the reflection of a point in the tangent lines to a curve, and the envelope of the perpendicular bisectors of the line segments between a given point and the points on a curve.

Let F_1 and F_2 be the two foci of an ellipse E. The Bivens transform of an ellipse moves each point A of E away from F_2 by the reciprocal of the distance from A to F_1. That is, the point A is transformed to the point $C = F_2 + (A - F_2)/||A - F_1||$. We denote this transformation by $B(A) = C$. Figure 14 shows the geometry.

Since the foci for a circle are the same point, the Bivens transform of any circle is the unit circle. So the Bivens transform of an ellipse can be viewed as a two-point normalization of the ellipse. Exercise 2 asks the reader to show that the semi-minor axial length of $B(E)$ is always 1. Furthermore, the Bivens transform of an ellipse is another ellipse.

If E has eccentricty e, $B(E)$ is an ellipse with eccentricity $\dfrac{2e}{1 + e^2}$. (24)

To see this, let the equation of E be $r = f(\theta) = b/(1 + e\cos\theta)$, where b is a positive constant. The foci in rectangular coordinates are $F_2 = (0, 0)$ and $F_1 = (-2be/(1 - e^2), 0)$. Let $g(\theta) = f(\theta)(\cos\theta, \sin\theta)$. Let $A = g(\theta)$ be a point on the ellipse, where θ is any angle. The distance from F_2 to A is

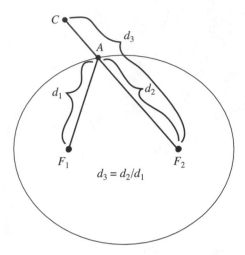

Figure 14. The Bivens transform

$f(\theta)$ and $||F_2 - A|| + ||F_1 - A|| = 2b/(1 - e^2)$. Thus the distance from F_1 to A is

$$||F_1 - A|| = \frac{2b}{1 - e^2} - f(\theta) = \frac{b(1 + e^2 + 2e\cos\theta)}{(1 - e^2)(1 + e\cos\theta)}.$$

Therefore the distance from F_2 to C is

$$r = \frac{f(\theta)}{||F_1 - A||} = \frac{b}{(1 + e\cos\theta)} \cdot \frac{(1 - e^2)(1 + e\cos\theta)}{b(1 + e^2 + 2e\cos\theta)}$$

$$= \frac{(1 - e^2)/(1 + e^2)}{1 + 2e\cos\theta/(1 + e^2)},$$

which is the polar equation for an ellipse with eccentricity $2e/(1 + e^2)$. Figure 15 shows the Bivens transform of two different ellipses. The dashed curves are the original ellipses and the solid curves are the transformed ellipses.

Given a curve Γ and a point A, the orthotomic of Γ is the reflection of A in the tangent lines to Γ. Figure 16 shows the orthotomic of the ellipse $r = 1/(1.5 + \cos\theta)$ where A is the origin. See Bruce [15] for a good discussion of orthotomics.

Another way to transform Γ with respect to a point A is to let $\mathcal{F}$ be the family of perpendicular bisectors of the segments with endpoints A and B, where B is any point on Γ. We refer to the envelope for $\mathcal{F}$ as the bisector curve. Figure 17 shows the bisector curve for the ellipse $r = 1/(1.5 + \cos\theta)$

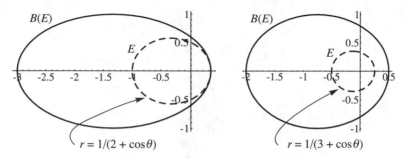

Figure 15. Bivens transforms of ellipses

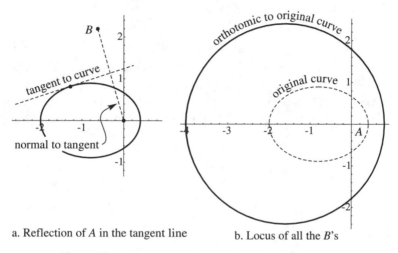

a. Reflection of A in the tangent line   b. Locus of all the B's

Figure 16. The orthotomic for an ellipse with respect to A

with respect to the focus F_1. The orthotomic of this bisector curve with respect to F_1 is the original ellipse, as shown in [**15**, p. 133]. Hence this fish-shaped curve is called the antiorthotomic of an ellipse. This curve looks suspiciously like the curves in Figure 13 where ω is at least as large as 8. Our goal is to show that these envelopes converge to the antiorthotomic as ω increases.

Interpreting a fishy envelope

As the rotation rate of the beacon increases, the shape of the envelope stabilizes, with only its scale changing. Figure 18 shows the envelope for $\omega = 50$ and for $\omega = 500$ corresponding to the lake $r = 1/(2 + \cos\theta)$. The envelope corresponding to $\omega = 500$ is one-tenth the size of the envelope corresponding

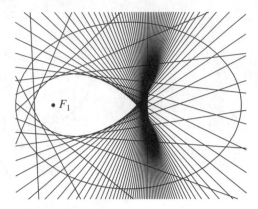

Figure 17. The antiorthotomic for an ellipse

to $\omega = 50$. Let $\rho = 1/\omega$. As ρ goes to 0, the fish-shaped envelope collapses to F_1. What this suggests is that once ω is at least as large as about 20 (for this lake and beacon), ρ is small enough so that the envelope generated corresponds to the first order term of the function generating it.

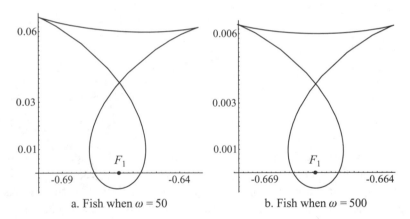

a. Fish when $\omega = 50$ b. Fish when $\omega = 500$

Figure 18. Scaled fish

Let us modify the functions in the Bivens parametrization to account for the variable $\rho = 1/\omega$. The functions g, $\mathbf{T}$, $\mathbf{N}$, and κ all remain unchanged. However, the ρ-analog to (12) becomes

$$\cos\alpha(\theta, \rho) = \frac{\rho + f'(\theta)}{\|g'(\theta)\|},\qquad(25)$$

(except when the right-hand side exceeds 1, which we ignore because we are interested in what happens at $\rho = 0$, in which case the right-hand side is bounded above by 1). Formulas (16) and (17) become

$$\mathbf{W}_1(\theta, \rho) = \cos\alpha(\theta, \rho)\,\mathbf{T}(\theta) + \sin\alpha(\theta, \rho)\,\mathbf{N}(\theta),$$
$$\mathbf{W}_2(\theta, \rho) = -\sin\alpha(\theta, \rho)\,\mathbf{T}(\theta) + \cos\alpha(\theta, \rho)\,\mathbf{N}(\theta), \qquad (26)$$

so

$$\frac{\partial \mathbf{W}_1}{\partial \rho} = \frac{\partial \alpha}{\partial \rho}\mathbf{W}_2, \quad \frac{\partial \mathbf{W}_1}{\partial \theta} = \left(\kappa\|g'\| + \frac{\partial \alpha}{\partial \theta}\right)\mathbf{W}_2,$$

$$\frac{\partial \mathbf{W}_2}{\partial \rho} = -\frac{\partial \alpha}{\partial \rho}\mathbf{W}_1, \quad \frac{\partial \mathbf{W}_2}{\partial \theta} = -\left(\kappa\|g'\| + \frac{\partial \alpha}{\partial \theta}\right)\mathbf{W}_1. \qquad (27)$$

Then, just as in (23),

$$P(\theta, \rho) = g(\theta) + Q(\theta, \rho)\mathbf{W}_1(\theta, \rho),$$

$$\text{where} \quad Q(\theta, \rho) = \frac{\|g'(\theta)\|\sin\alpha(\theta, \rho)}{\kappa(\theta)\|g'(\theta)\| + \dfrac{\partial \alpha}{\partial \theta}(\theta, \rho)}. \qquad (28)$$

By (28) and (27),

$$\frac{\partial P}{\partial \rho} = \frac{\partial Q}{\partial \rho}\mathbf{W}_1 + Q\frac{\partial \alpha}{\partial \rho}\mathbf{W}_2. \qquad (29)$$

Thus, when ρ is small,

$$P(\theta, \rho) \approx F_1 + \rho\,\frac{\partial P}{\partial \rho}(\theta, 0).$$

This is the first order Taylor approximation for P with respect to ρ. What we shall show is that the curve $\frac{\partial P}{\partial \rho}(\theta, 0)$ is an antiorthotomic.

By (29) and (27), the tangent vector to $\frac{\partial P}{\partial \rho}(\theta, \rho)$ with respect to θ is

$$\frac{\partial^2 \mathbf{P}}{\partial \theta \partial \rho} = \left(\frac{\partial^2 Q}{\partial \theta \partial \rho} - Q\frac{\partial \alpha}{\partial \rho}\left(\kappa\|g'\| + \frac{\partial \alpha}{\partial \theta}\right)\right)\mathbf{W}_1$$

$$+ \left(\frac{\partial Q}{\partial \rho}\left(\kappa\|g'\| + \frac{\partial \alpha}{\partial \theta}\right) + Q\frac{\partial^2 \alpha}{\partial \theta \partial \rho} + \frac{\partial Q}{\partial \theta}\frac{\partial \alpha}{\partial \rho}\right)\mathbf{W}_2. \qquad (30)$$

When $\rho = 0$, the coefficient of $\mathbf{W}_2$ in (30) is 0. To see this, when $\rho = 0$, P is the focus F_1, $\mathbf{W}_1(\theta, 0)$ points in the direction from $g(\theta)$ to F_1, and $Q(\theta, 0)$

is the distance from $g(\theta)$ to F_1. So,

$$F_1 = P(\theta, 0) = g(\theta) + Q(\theta, 0)\mathbf{W}_1(\theta, 0),$$

from which

$$\mathbf{0} = \mathbf{T}\|g'\| + \frac{\partial Q}{\partial \theta}(\theta, 0)\mathbf{W}_1(\theta, 0) + \left(\kappa\|g'\| + \frac{\partial \alpha}{\partial \theta}(\theta, 0)\right)\mathbf{W}_2(\theta, 0),$$

which by the ρ-modified analog to (21) is

$$\mathbf{0} = \left(\|g'\|\cos\alpha(\theta, 0) + \frac{\partial Q}{\partial \theta}(\theta, 0)\right)\mathbf{W}_1(\theta, 0)$$

$$+ \left(\kappa\|g'\| + \frac{\partial \alpha}{\partial \theta}(\theta, 0) - \|g'\|\sin\alpha(\theta, 0)\right)\mathbf{W}_2(\theta, 0).$$

Thus

$$\frac{\partial Q}{\partial \theta}(\theta, 0) = -\|g'\|\cos\theta(\theta, 0). \tag{31}$$

From (28),

$$\left(\kappa\|g'\| + \frac{\partial \alpha}{\partial \theta}(\theta, \phi)\right)Q(\theta, \phi) - \|g'\|\sin\alpha(\theta, \phi) = 0. \tag{32}$$

Differentiating (32) with respect to ϕ and then putting $\phi = 0$ gives

$$\left(\kappa\|g'\| + \frac{\partial \alpha}{\partial \theta}(\theta, 0)\right)\frac{\partial Q}{\partial \rho}(\theta, 0)$$

$$+ Q(\theta, 0)\frac{\partial^2 \alpha}{\partial \theta \partial \phi}(\theta, 0) - \|g'\|\cos\alpha(\theta, 0)\frac{\partial \alpha}{\partial \rho}(\theta, 0) = 0. \tag{33}$$

In (33), we can, by (31), replace $-\|g'\|\cos\alpha(\theta, 0)$ with $\frac{\partial Q}{\partial \theta}(\theta, 0)$, which shows that the coefficient of $\mathbf{W}_2$ in (30) is 0 when $\rho = 0$. So we have shown that the tangent to $\frac{\partial P}{\partial \rho}(\theta, 0)$ at any θ_0 is parallel to the direction from $g(\theta_0)$ to F_1. Thus $F_1 + \frac{\partial P}{\partial \rho}(\theta, 0)$ is the envelope of the family of lines $\mathcal{F} = \{M(\theta)| \theta \in \mathcal{R}\}$ where $M(\theta)$ is the line through the point $F_1 + \frac{\partial P}{\partial \rho}(\theta, 0)$ whose direction is from $g(\theta)$ to F_1 (which is the same direction as $\mathbf{W}_1(\theta, 0)$). See Figure 19 for the geometry.

Now reflect F_1 through each line of $\mathcal{F}$. Since $\mathbf{W}_1(\theta, 0)$ is parallel to $M(\theta)$, and since $\frac{\partial P}{\partial \rho} = \frac{\partial Q}{\partial \rho}\mathbf{W}_1 + Q\frac{\partial \alpha}{\partial \rho}\mathbf{W}_2$ (which is (29)), $M(\theta)$ is $|Q(\theta, 0)\frac{\partial \alpha}{\partial \rho}(\theta, 0)|$

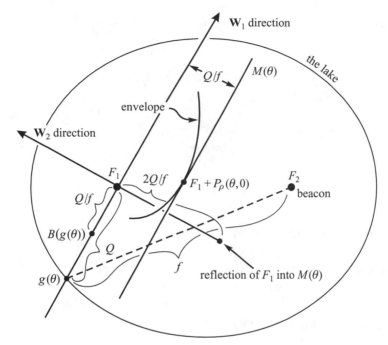

Figure 19. The geometry of the curve $F_1 + \frac{\partial P}{\partial \rho}(\theta, 0)$, where $P_\rho = \frac{\partial P}{\partial \rho}$

from F_1. By (25), $\cos \alpha(\theta, 0) = f'(\theta)/\|g'(\theta)\|$, so

$$\sin \alpha(\theta, 0) = \frac{f(\theta)}{\|g'(\theta)\|}. \tag{34}$$

Differentiating (25) with respect to ρ gives $-\sin \alpha \frac{\partial \alpha}{\partial \rho} = 1/\|g'\|$. Solving for $\frac{\partial \alpha}{\partial \rho}$, evaluating at $\rho = 0$, and using (34) yields

$$\frac{\partial \alpha}{\partial \rho}(\theta, 0) = -\frac{1}{\sin \alpha(\theta, 0)\|g'(\theta)\|} = -\frac{1}{f(\theta)}. \tag{35}$$

Therefore the distance from F_1 to $M(\theta)$ is $Q(\theta, 0)/f(\theta)$. The Bivens transform of $g(\theta)$, $B(g(\theta))$, is the point $g(\theta)$ moved away from F_1 by the reciprocal of the distance from F_2 to $g(\theta)$. That is, the distance from $B(g(\theta))$ to F_1 is $Q(\theta, 0)/f(\theta)$. By (35), $Q(\theta, 0)\frac{\partial \alpha}{\partial \rho}(\theta, 0)$ is negative, so by (29) the line through O in the $\mathbf{W}_1(\theta, 0)$ direction separates the point $\frac{\partial P}{\partial \rho}(\theta, 0)$ and the ray $\mathbf{W}_2(\theta, 0)$. Thus, if F_1 is reflected in $M(\theta)$, it ends up at the same place as it would by scaling $g(\theta)$ away from F_1 by a factor of $2/f(\theta)$ and then rotating $90°$ counterclockwise. That is, the orthotomic of $\frac{\partial P}{\partial \rho}(\theta, 0)$ is

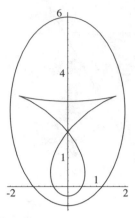

Figure 20. $\frac{\partial P}{\partial \rho}(\theta, 0)$ is the antiorthotomic of an ellipse

an ellipse obtained by magnifying the Bivens transform of the elliptical lake
by a factor of 2 and then rotating a quarter turn counterclockwise about F_1.
Figure 20 shows $\frac{\partial P}{\partial \rho}(\theta, 0)$ and its orthotomic ellipse. Since the orthotomic
of the curve $\frac{\partial P}{\partial \rho}(\theta, 0)$ is an ellipse, then $\frac{\partial P}{\partial \rho}(\theta, 0)$ is the antiorthotomic of an
ellipse.

A summary

In this chapter we examined a beginning calculus exercise whose standard solution has puzzling implications. Improving and generalizing the underlying
model so as to account for finite light speed, convex lakes, and a spectator's
vantage point led to the notion of ordinary points, the places where, for a
given rotation rate, a spectator will always see the beacon spot on the beach
move counterclockwise along the beach. By pressing this definition of ordinariness, we improved our ability to portray the ordinary points of the lake
for any rotation rate of the beacon, ultimately leading to the definition of a
family of separating lines.

The primary idea used in deriving a parametrization for the envelope for
this family was that of a dynamic coordinate system. We used the tangent and
normal coordinate system whose hub moves along the shoreline of the lake,
and the separating line and its normal, W_1 and W_2, whose hub we sometimes
took as being on the beach or at the non-beacon focus. This same approach of
exploiting a good choice of perspective has been used throughout this book.
Heavenly motion as induced by mass attraction is characterized by using the
$\mathbf{u}_r$ and $\mathbf{u}_\theta$ dynamic coordinate system that moves along with the falling body.

The dynamic perspective is a key analytical tool. With its aid, we caught a fish (the antiorthotomic of the Bivens transformation of an elliptical lake) as we watched the beacon sweep along the beach.

Since light travels so quickly, the chances of actually viewing the phenomenon discussed in this chapter are slim, at least for a terrestrial lake. However, consider an astronomical situation. Imagine a pulsar with a single bright spot on its equator rotating with a period of seconds, and that light from this bright spot is emitted as a laser shooting off into space. Imagine also that a dense asteroid belt circles the pulsar in an elliptical path in its equatorial plane with the pulsar at one focus of the ellipse. The laser beam from the pulsar sweeps along the asteroid belt. Where can we position a spaceship from which the light of the pulsar's beacon illuminating the asteroid belt always appears to be going counterclockwise? The answer, as we have seen, is near the other focus of the ellipse. Someday, if the science fiction vision of man reaching the stars becomes reality, someone may witness this very phenomenon.

Exercises

1. For a parametrization $g(\theta) = (x(\theta), y(\theta))$ of a curve, the unit tangent and normal are $\mathbf{T}(\theta) = (x'(\theta), y'(\theta))/\|g'(\theta)\|$ and $\mathbf{N}(\theta) = (-y'(\theta), x'(\theta))/\|g'(\theta)\|$. Show that $\mathbf{T}'(\theta)$ is given by

$$\mathbf{T}'(\theta) = \frac{\|g'\|^2(x'', y'') - (x', y')(x'x'' + y'y'')}{\|g'\|^3}$$

$$= \frac{x'y'' - x''y'}{\|g'\|^3}(-y, x) = \kappa \|g'\| \, \mathbf{N},$$

where κ is given by (18). Similarly, show that $\mathbf{N}' = -\kappa \|g'\| \mathbf{T}$.

2. Show that the length of the semi-minor axis of the Bivens transform of an ellipse (24) is 1.

3. Determine whether the orthotomic of an ellipse is an ellipse. If so, find its eccentricity in terms of the eccentricity of the original ellipse.

4. Generate the envelope for the rotating beacon problem with the beacon at the origin and a lake having the polar equation $r = 1/(3 + \cos\theta)$.

5. Find the coordinates where the y-axis intersects $\frac{\partial P}{\partial \rho}(\theta, 0)$ in Figure 20. Find the coordinates of the cusp on the tail of the fish.

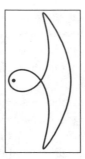

Figure 21. The second Taylor term

6. The graph of $\frac{\partial^2 P}{\partial \rho^2}(\theta, 0)$ as shown in Figure 21 looks like a clothes hanger. Generate the graph of $\frac{\partial^3 P}{\partial \rho^3}(\theta, 0)$.

7. Generate the orthotomic of the curve in Figure 21 with respect to the given point (which is the origin).

8. Let a nonconvex shoreline be given by the polar equation $r = 1 + \cos\theta$, a cardioid, with the beacon at the origin. Although this curve and beacon position violates some of the assumptions made in this chapter, generate the corresponding envelope for various rotation rates ω. Contrast the resulting envelopes with those of Figure 10, which corresponds to a circular lake with the beacon at a non-focus point.

9. Find the Bivens parametrization of the bungee cord curve of Chapter X. Let $f(\theta) = 0$, $g(\theta) = (\cos\theta, \sin\theta)$, and let $\mathcal{F}$ be the family of chords with endpoints $g(\theta)$ and $g(p\theta/q)$, where p and q, $p > q$, are positive integers. Figure 22b shows the family when $p = 9$ and $q = 4$. $\mathbf{T}(\theta)$ makes an angle of $\pi/2 + \theta$ with the horizontal. Since the central angle between $g(\theta)$ and $g(p\theta/q)$ is $p\theta/q - \theta$, we have $\alpha(\theta) = (p-q)\theta/q$, as indicated in Figure 22a. Then $\mathbf{W}_1(\theta)$ makes an angle of $\pi/2 + (p+q)\theta/q$ with the horizontal, so

$$\mathbf{W}_1(\theta) = \left(-\sin\frac{(p+q)\theta}{q}, \cos\frac{(p+q)\theta}{q} \right).$$

Show that $Q(\theta) = 2q\sin((p-q)\theta/(2q))/(p+q)$ and that

$$P(\theta) = \left(\frac{p\cos\theta + q\cos\left(\frac{p\theta}{q}\right)}{p+q}, \frac{p\sin\theta + q\sin\left(\frac{p\theta}{q}\right)}{p+q} \right).$$

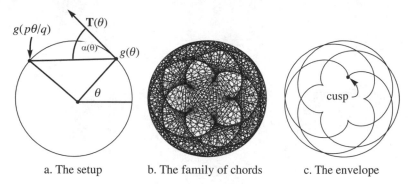

a. The setup   b. The family of chords   c. The envelope

Figure 22. The bungee cord problem, $p = 9$, $q = 4$

10. Recall that $P'(\theta)$ points in the $\mathbf{W}_1(\theta)$ direction so that $P' = (Q' + \cos\alpha)\mathbf{W}_1$. Cusps appear on the envelope when $P'(\theta) = \mathbf{0}$. Use the parametrization in Exercise 9, to find a formula in terms of p and q for the number of cusps in an epicycloid. For $p = 9$ and $q = 4$, the envelope has five cusps as shown in Figure 22.

preamble xiv

But true love is a durable fire
From itself never turning.
Sir Walter Relegh (1552–1618), *As You Came from the Holy Land*

A few years ago, I visited the Yucatan for a short course on ancient Mayan mathematics. We heard lectures, were transported long distances by bus, and walked among the ruins under a hot sun. We also climbed pyramids. Climbing up sixty steeply slanted steps is the easy part. Other pyramids may have 91 steps or 360 steps, or some other metaphysically important Mayan number, leading to greater heights. At the top, it's over six stories high, just the climber and open space, towering over the scrub brush of the jungle. Going down is the hard part. The steps are much more worn and impossibly narrow than they were moments ago, on the upward climb. The oppressive heat abruptly vanishes, an unwelcome coldness grips the spirit.

The Mayans were great builders and careful observers of the sky. Much of their history has vanished in mystery. That being the case, I have borrowed some of this mystique, their obsession with the movements of the planets, their tradition of calendars and cycles, and have inflated their already impressive engineering skills—to use as background.

This next chapter is a story. It incorporates all the aspects of falling and spinning as presented in this book. Furthermore, it does so in the spirit of Jules Verne and H.G. Wells against a Mayan backdrop. The story can be viewed as a long word problem. At its end is a series of ten exercises, serving as a final examination of the material in this book.

In the story, much is made of the Mayan long count, a calendar whose cyclical length is 5126 years and which is slated to roll over on the winter solstice of the Gregorian year 2012, a roll-over time signaling uncertain doom. The Mayans used base 20—they counted on their toes as well as their fingers. Every day in this 5126 year period has a unique number attached to it, represented as the five-tuple $p.q.r.s.t$, where t is any of the twenty days of a week; s is any of the eighteen weeks of a 360 day year or *tun*; r is any up to 20 tuns; q, the *katun*, is any up to 20 scores of tuns; finally, p is any up through 12 scores of katuns. Thus the number 8.2.4.1.3 represents day $8(20 * 20 * 360) + 2(20 * 360) + 4(360) + 1(20) + 3 = 1,167,863$. The day before the roll-over is 12.19.19.17.19. The roll-over day represents 5200 tuns. It is also 5126 cycles of the earth about the sun. Therefore, if x represents the number of days in earth's transit about the sun, then $5126x = 5200 * 360$, a Mayan way of saying that an earth year is about $x \approx 365.2$ days.

CHAPTER **XIV**

The Long Count

A small yellow fish with a thin blue stripe from jaw to tail lingered outside the viewport. Except for the thick glass portal and a few antennae, the hull of the ship matched the fish's color. We were headed for the bottom, six hundred feet down. Priscilla's buoyancy was mildly negative. She sank slowly along a vertical wall of colorful coral.

Sarah snuffled up to the window. Her wet nose pressed against it. She was the ship's mascot for the day. We wouldn't be down long. I rubbed her ears, which was what she wanted all along.

As we passed two hundred feet, the fish left, racing off to a protective yellowish splotch on the wall. So too had Sarah. She lay at my feet in a ball, her nose on her tail. I wiped the window clean. In another fifteen minutes, we settled on the sandy bottom. The shapes outside were blue shadows of varied darkness.

The year was 2012. We were on an archeological dig, hopefully to unearth Mayan artifacts. The Mayans had anticipated the end of the world on the winter solstice of this year—just a few days from now—the end to their 5126 year count-down to dissolution. Maybe what we brought to the surface could add some color to their otherwise gloomy forecast.

"Okay, Bill, let her go," I phoned to Titus, the mother ship. In twenty seconds a puff of sand exploded from the ocean floor. Bill had dropped a ten pound ball of diameter four inches, a homing beacon inside.

"Good shot," I said, and asked for the next one.

The next two trials were drops of twenty and fifty pound balls with descent times of 17 and 12 seconds. I had always wanted to check the drag formulas of college physics days. I also needed practice on search and retrieval. Not every middle-aged archaeologist gets to solo pilot a submersible. This was a dream come true.

The first two balls were easy to retrieve. The sub's mechanical arm plucked them like cucumbers from a vine. As I worked at the last ball, yellow flashes caught my eye. We were off the Yucatan. A Spanish galleon from the sixteenth century had sunk here. We had already recovered its anchor and some cannon.

Heart beating loudly in my ears, sweat beading on my brow—gold for the taking will do that to a man—I manipulated the claws of the arm and extracted three yellow disks. In moments, they passed through Priscilla's pressure lock.

I bent to look, and blew away dirt specks.

Coins! Our luck was changing. The face image was a disk behind a triangle. I reached for a coin, to feel its weight and to see it more clearly.

That's when it happened.

The sea floor opened like a zipper on an over-stuffed suitcase. The sandy floor spilled upwards as water cascaded downwards. Inside Priscilla, Sarah and I were knocked back and forth. I struggled for a seat and grasped its base pole. My chin pressed into its cushion.

We were in free fall. All sense of weight vanished. I strapped myself in. Sarah floated in midair. I grabbed her by the tail and pulled her onto my lap. Nonsense sounds and images flooded my mind. A childhood refrain played repeatedly:

There's a hole in the bottom of the sea.

What would happen if the plug were removed? Would all the oceans vanish as bathwater down the drain or as down the throat of Thor, a Norse god, who once tried to drink the oceans dry?

We fell for three minutes.

Priscilla splashed into a deep reservoir, the crash of the accompanying cascade of waters cushioning the impact. The submarine's descent slowed, stopped, and reversed.

We were somewhere between ten and a hundred miles beneath the earth's surface. Priscilla was designed to withstand ocean pressures up to three miles deep. Her hull held. The only conclusion was that we were in an underground sea, and the crevice that had opened had sealed itself. What chance did we have to get back? I stroked Sarah. If Priscilla's engines yet worked, we could surface for a look around before Davy Jones came for us.

Soon Priscilla's smooth pumping action emptied ballast tanks and stead-
ied my nerves. I thought back to the sunshine. I hoped Titus had weathered the quake. The treasure we sought was gone, down at the bottom of
this underground sea. Except for three coins scattered on the submarine
floor. Sarah helped me find them, picking them up with her tongue and
teeth. I wiped the saliva-covered gold pieces on my shirt. The coins were
identical.

Figure 1. Earlobe plugs

A close look revealed the disk to be the moon. The pyramid was Mayan.
Around the coin's edge were glyphs, looking like misshapen dwarves with
square heads and feet. On the coin's flipside was a pendulum, below which
was a stylized eye. The rims of the coins were beveled inwards to a spool
shape. If they were old Mayan, they probably functioned more as jewelry
than as money. They were worn as decorative plugs in pierced ear lobes,
and could be traded for goods and services. In that limited sense, they were
coins.

At the surface, I opened the hatch. Priscilla was cigar-shaped with a diameter of eight feet and no conning tower. I stood on a ladder, my arms resting
on her smooth hull. The air was heavy yet breathable with a trace of sulphur.
Down below, Sarah sneezed. She lay on the floor, both paws covering her
eyes and muzzle.

As my eyes adjusted from the lights in the submarine, a greenish
phosphorescence sparkled from the water's surface and from the rocky

canopy overhead. We were in a current, and were drifting along at several knots.

I took inventory. The galley contained enough food for a month. If the atmosphere worsened, we had air for a week. Toiletry facilities were primitive—a flush system dumped waste directly into the sea. There was no shower. Finding fresh water was a must. We had two gallons.

I tried the radio. No answer. I hadn't expected any. Titus must think we were lost. Ironically, the doom the Mayans prophesied had come true in the earthquake, albeit a few days early.

For a day we went with the current. I rigged an outside platform for Sarah so she could see where we were going. A whale drifted beside us, seeking company. Sarah barked, her favorite activity. At that, the whale dove, its flukes splashing her. It must have gotten sucked down in the quake with us.

Priscilla's sonar scopes finally registered a bottom. After that, the sea floor rose rapidly. The current quickened through a narrow chasm. A waterfall ahead meant disaster. Using a spotlight, I searched the shoreline on either side, seeking some place to tie up. The river split.

Which way to go? I gambled right.

The right-hand branch calmed into a smooth lake with a gently sloping shoreline. Ahead in the semi-gloom of the spotlight was what appeared to be a dock. A natural promontory, my mind insisted. I moored the submarine. Sarah jumped ashore, barking, the echoes ricocheting from all sides. There's no one here. Let her bark, I assured myself.

A pack on my back, a torch in hand, I explored. Sarah ran circles around me and dashed off, returned and barked louder, more excitedly. Her muzzle was dripping. She had found fresh water. A spring trickled from the walls. A small creek drained into the lake. Making many trips, I filled all the containers I could find.

During the trips back and forth, I studied the ground. Steps had been cut into the earth. Tunnels branched off from the beach. I followed one for a hundred yards. In the semi-darkness, I stumbled against some flotsam, and kicked a container. A canteen. On its face were initials, JV.

Jules Verne! It was rumored that his fantasies were slight fictional accounts of his own adventures. But in *Journey to the Center of the Earth,* Verne had been a hundred miles below Iceland, not the Caribbean. If this canteen was his, it had floated through a maze of interconnecting passageways, as it could if Socrates' vision of an underground river network was accurate. I kept it for luck.

Figure 2. The elevator frames

Down a second tunnel, carvings adorned the walls. They seemed familiar. Pendulums, moons, and pyramids. The coins. I retrieved one from the sub. The carvings were the same icons, the same style. My excitement grew. Before me was a door. I could see the joints. "Open sesame," I muttered. There was no doorknob. Instead there was a slot the same size as the coin. I hesitated. But they were of little value if I kept them in my hand.

Inserting one of the coins, I offered a prayer, and hoped.

The door swung open.

Inside was a room of great height bathed in an orange glow. Dangling from the ceiling, a chain swept back and forth, its free end nearly touching the ground. On the floor was a mosaic dial. A hairy face stared blankly from its center, a thick snakelike tongue protruding upwards from the plane of the floor. This place didn't seem to be abandoned.

Around the room's walls were painted carvings. The first frame showed a face with no protruding tongue. In the next frame was a box containing smaller boxes, clearly inside a circle, perhaps a glyph representing earth. The third frame showed a pendulum aligned with the main spoke of the wheel, a conjunction. The next frame showed a fire beneath the box. The fifth showed the same box of smaller boxes outside the circle. The last showed the first image with its tongue protuding.

What could it mean? Surely I should be able to read the message after years of study.

"Senor, you wish to go up?"

I froze. A human voice. A strange Spanish dialect.

I turned. Before me stood a short, thin figure. A man!

Madre de Dios!

Hallucination? I approached the spectre. He stood his ground, looking at me with steady eyes. I grasped his shoulders. Solid flesh. I embraced him and wept.

Sarah had been strangely silent. Her tail was down. She couldn't make up her mind whether to bark or cringe.

"Who are you?" I asked, slipping into Mayan. His earlobes bore the same tokens as my coins. From his wrists dangled jade bracelets. His hair was braided with care. His face was worn. He had to be twice my age.

At my question, he smiled. "Ah, the tongue. It is almost time. We were hoping that someone like you would come."

I fired a barrage of questions. He patiently fielded them, explaining that long ago, when the Spaniards arrived, systematically destroying their culture, tearing down their temples, burning their codices, and all looked lost, the high priests and their families sealed themselves within these caves and mines, vowing to carry on until the long count ended. Buried in their refuge, according to their visions and prophecies, they hoped to usher in a new age. That was their dream.

"We sealed ourselves in too well. The old excavations collapsed. We work at new tunnels to the surface. The task is great and we are few."

I had thousands of questions. He knew the answers to many of the puzzles of my research. Now was not the time to ask them.

He was patient with me, and explained that his community was far away. He came at regular intervals to tend the pendulum, to ensure that all was in readiness for the great day.

"Yes, the long count," I understood. Some final rites needed to be performed before time ended. "What is the ceremony?"

He waved at the wall. "You wish to go up?" He repeated the question by which he had greeted me. He explained that when the tongue of the image was depressed, the mechanism began preparing itself. At that he touched the tongue, and it receded smoothly into the floor. "There will be one last journey from this underworld via the offering."

A chill went through me as he said those words.

He moved deliberately, never in a hurry, sure of his purpose.

"Come, let me show you."

We left the room. He continued to speak. What he described, the image sequence on the wall, was an elevator to the surface of the earth.

"Why haven't you used it before?"

He waved both of his arms. "All of this will be damaged when we use it. We do not have the manpower to ready it a second time. We've saved it for the last day. It is our destiny!"

He labored over the last phrase.

After a pause, he returned to his graceful style, and added, "Neither can we walk the passageway. Parts of it get too hot."

As I mulled over his words, he asked, "Your ship, may I see her?"

I shook off a twinge. There was no reason not to let him on board. He had probably already inspected her from the outside and looked inside. I'd left the hatch open.

The shoreline had changed. Depressing the tongue in the pendulum room had initiated activity. A sluiceway, invisible before, had opened not far from the dock.

"In there," he motioned down the tunneled waterway, "is the portal to the surface. Shall we go?"

I helped him onto the submarine. With Priscilla's lights ablaze, we sailed into the dungeon. After a mile the canal petered out.

"This is it," he said. We stood on the hull of the ship surveying the chamber. "In a little while, the earth will erupt in controlled fashion. Steam jets will spring forth all about the tunnel, and this," he kicked Priscilla, "shall rise."

A chest of gray metal bound by brass bands stood on the shore. We waded over to it.

"These are the best of our treasures," he explained. He extracted a token from his ear lobe, and inserted it into a slot in the chest. The heavy lid eased open. The priest removed a vessel containing clear liquid. Chanting, he sprinkled it in all directions. He poured some onto a cupped palm, and drank. He indicated to me to do likewise.

Caught up in the spirit of the ceremony, I held out a hand for a few drops and sipped. It was sweet and bitter.

He took out a bundle of purple cloth. Inside was a ceremonial dagger, its handle crusted with jewels. He slit a wrist. Drops of blood dabbled the pool of water. He gestured to my arm.

I declined.

"It is enough," he nodded, and closed the lid, but not before retrieving his token and reinserting it in his ear. "Take this." He wiped the blade clean, and handed it to me. "You will go up." He looked at me keenly, then closed his eyes, and went off chanting. The chest receded into the floor.

I felt woozy. Smoke began filling the chamber. I needed to sit. Clambering down Priscilla's hatch, I forced Sarah into a padded box, and strapped myself into a chair. I was almost asleep. Dimly, I knew it had to be the ceremonial drink. Dumb! Is everyone as nearsighted as I? My limbs lost their mobility. I fought to keep my eyes open.

A gate descended, sealing the chamber. The waters, which had been a dull black, were now an oily mixture. Priscilla trembled. The oil bath bubbled. The temperature rose. Slowly at first, the submarine moved up the ramp. Sprays emanated from all around. The diameter of the tunnel shrank. A roar

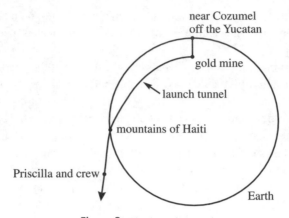

Figure 3. The launch tunnel

echoed through the earth. Our speed increased. In controlled experiments, some men have briefly endured forces forty-five times that of surface gravity. The maximum survivable sustained acceleration is about 10 g. I was pressed back into my seat. Sarah growled. She couldn't bark. I couldn't move my arms. My tongue slipped back against my throat. We rocketed along the tunnel. Vision faded and I lost consciousness.

When I came to, I felt light. Down was now any direction away from the central axis of Priscilla's longitudinal body. I peered outside. Stars swirled in dizzying fashion.

We were in space, rotating once every six seconds.

I tried reconstructing what had occurred from what we knew of the Mayans and what the old priest had said. I knew that in ceremonies, priests tore beating hearts from prisoners of war. "Lord, he could have taken my heart," I thought, and glanced at the blade the priest had given me. The Mayans had erected mammoth pyramids, destroying countless lives in the process. They were great architects and careful astronomers. Their calendars were more precise than those of the Greeks and Romans. They had been wealthy, although not as rich as the Aztecs and Incas. I now knew that their gold was mined from deep within the earth. The planets were their gods. Evidently, the priesthood sent sacrifices to the stars.

Labor had been no problem. There were always prisoners of war. From their mines, they chiseled a smooth tunnel, arcing for hundreds of miles, exiting in the mountains of Haiti, as I learned later and as shown in Figure 3. Through experimentation, tapping into volcanic power and steam repositories, whose energy was released in careful synchronization along the tunnel, they were able to throw their sacrifices so fast that they never came down.

This time Priscilla was the offering. I had hoped that the elevator went to the earth's surface. It had done much more than that.

Adrift in space, I sorted through my emotions. I wondered whether the priest had sacrificed himself, or whether he escaped in time to return to his family. He had known the elevator would blast me into oblivion. He must really believe. What must it take for a community to survive 450 years in hell, and live to fulfill their destiny as they saw it?

In an effort to force my mind to think about survival strategies, I tried calming myself by flipping a coin. After a few trials, I could toss it back to myself, allowing it to spin in multiple loops before snatching it out of the air.

Sarah and I were in good shape, except for a few bruises. The submarine withstood the absence of outside pressure. I began to reduce the air pressure in Priscilla, getting our bodies back to normal conditions.

Except for Priscilla's operating manual, the only reading material on board was Bill's water-stained copy of H.G. Wells' *The First Men in the Moon.* It seemed almost like my own story. As for a gravity blocking substance, I shook my head. "Cavorite," I scoffed.

By now we had been in space for two days. I rigged a hand grip near the window and rotated myself in the direction opposite to Priscilla's spin, kicking against the side of the submarine. While Priscilla spun about me, I looked at the stars and where we were going. Off to one side, the moon loomed large. I stared for long minutes. A sinking realization came over me.

The pendulum mechanism in the mines had been synchronized with the moon. The Mayans were much better engineers than I had reckoned. Their sacrificial offerings not only were aimed at the moon, they literally hit the moon. We didn't have long to impact.

What do you do with a day, knowing it's your last one? The condemned man gets a last meal or a last smoke. I looked in the larder. Beans and rice it was. I opened a can of ham for Sarah. We splurged on water. I took a sponge bath. I tried the radio again. No response.

Periodically I looked at the moon and patted Sarah.

Somehow I fell asleep. I didn't intend to. I wanted to savor each moment left.

When I awoke, the moon filled the viewport. Knowing it was useless, I nevertheless placed Sarah in her pillowed cage. I strapped myself in. What would death feel like?

The moon rushed to greet us. We were heading into the center of a crater, like a dart towards a bullseye.

Bam.

I flinched.

We ripped through the surface.

It was like a spider web.

I glimpsed a huge funnel-shaped hole. We raced along its central axis. We tore down its shaft as in Figure 4. Priscilla's antennae was ripped away. She encountered controlled resistance, a collapsible network, but nothing solid. We slowed. In five minutes we stopped.

We were alive.

How? Something had covered the crater, but not man. The Mayans had been throwing artifacts for hundreds of years with the same accuracy that Priscilla had hit the moon. As I later learned, the moon had been inhabited by intelligent beings—Wells called them Selenites, after the Greek moon goddess, Selene. The Selenites had figured a way to receive their gifts from space without them being smashed on the moon's surface.

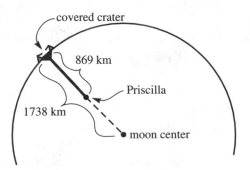

Figure 4. The lunar landing tunnel

The shaft was straight down. Priscilla had come to rest halfway to the moon's center, I estimated. While the surface of the moon has no atmosphere, the air pressure at the landing site was like being five miles high on the earth. I could breath for short periods of time. On forays, I carried a tank on my back.

The region was honeycombed with passages. During the moon's fortnightly-long day, an intricate system of Selenite mirrors directed sunlight well. But once the sun was gone, the long night would be very dark.

I found no trace of living Selenites.

But I now knew that Wells, like Verne, had written autobiographically. Wells had been an astronaut in *The First Men*. In that story, Mr. Cavor had been marooned. Maybe he had walked where I now stood.

Sarah and I explored the corridors. There were thousands of miles of tunnels in SeleniteLand. On these walls, the Selenites had left pictograms. Some passageways bore images of mushroom-like stalks, signs which led to what had been cafeterias. There were pictograms for going up and going down.

The graffiti was almost readable. The symbols were suggestive in nature—a ladder pictogram for stairs, a trellis pictogram for suspended walkways, a solid dot pictogram warning of pits in the passageways.

Then I spotted a Mayan symbol. I followed that sign along the passageways. They led to a wide hallway in which were displayed broken stone carvings of square-headed idols with holes in their bellies, jade necklaces, wooden dolls, and golden earrings. Within a glass case, a codex lay open, a book of folded, thin wooden panels with gaudily painted glyphs, illustrated with awkwardly posed human forms clad in quetzal feathers.

Some pages were catalogs of spices or herbs. The glyphs detailed recipes and potions. On one panel, folded and tucked inside a slit was a sheet of yellowed paper. There could be only one author, for Wells had reported that the Selenites had nothing like paper.

Fellow traveler,

Greetings from the past. I've put notes in exhibits that the historian caste delight in maintaining. They've allowed me to handle treasures like this document, presumably from the Middle Americas. They say that it fell from space! The last instance occurred over three hundred years ago. It seems authentic. I have prepared some notes for the Royal Society in London. You'll find them in the central strand. Look for the tower. I include a map.

James T. Cavor
Christmas Day, 1903

Sarah and I found the central strand. It was a great elliptical cavern, two miles across. Near one end stood a truncated pyramid, one hundred feet tall. A steep staircase was chiseled into each slanting face. The top of the pyramid was a level square, fifty feet to a side. In the center of this platform stood a smaller pyramid, made of metal, atop which sat a crystal cylinder. It looked like a lighthouse. One face was inscribed with glyph fragments. Another was covered in Selenite pictograms. A third contained mathematical formulas, being a hybrid of conventional mathematics and Selenite symbols. The fourth was in English:

In honor of his eminence, the Grand Lunar, the Mathematics Guild dedicate this Dance of Light. Only he will clearly see, all others catching shadows and reflections. In conjunction with the equinox and his touch, the dance begins.

Conjunction must mean the lunar noon, probably a special one during the course of the year. I only had one chance at this. Lunar noon was minutes away, as near as I could calculate. *His touch* could mean anything. During the ceremony, where would the Grand Lunar have sat?

Looking around, I saw various structures. His seat may have been the highest. The best candidate was a mile away. In this amphitheater with the pyramid at a focus, that hill seemed positioned at the other focus.

Descending the steep steps was more difficult than climbing. Sarah literally slid down, her legs flailing, her claws scraping for support, falling in slow motion. At the bottom she shook herself, as if emptying a coatful of water, and barked for me to hurry. I took my time, and backed down on all fours.

Once on the ground, to cover the distance, I took long leaps. Minutes later we were at the top of the Grand Lunar's observation point. A massive mound occupied most of the space. An evil smell swept over me.

Sarah growled.

It grumbled.

The heap was alive!

I forced myself forward. I shone a light over the mass. Smooth carapace and a pair of eyes reflected the glow. The eyes blinked.

"Welcome, earthman!"

The voice was low, and sounded as if it came from everywhere, deep, yet scratchy.

I looked about me. There were no others.

I stared at the wreck before me.

Then I remembered my manners. This must be the Grand Lunar.

I knelt, positioning myself lower than his eyes. Sarah curled beside me.

"Forgive me. I didn't think there was anyone here," I said stupidly. "May I get you anything?" He was obviously ill. I reached out to touch, and then thought better of it.

"The human said you would come. I have been waiting."

In the presence of death, I had no words.

"Yes, there is one thing. Once more." His tiny arms, dwarfed by a mammoth brain case, gestured at the floor. "I can no longer move."

The floor was almost the twin of the pendulum room floor back on earth. There was no open mouth, merely a slit, just big enough to accept a token. A small pile of them stood nearby.

I dropped one into the slot. Five seconds later came a faint echo. I looked around expectantly.

From atop the great pyramid, a light flashed. The crystal cylinder began rotating. A beam cut like a laser towards the periphery of the room. It flashed in wild color combinations. Around the room, images appeared. As the cylinder spun faster, these images shrank in size. I recognized Mayan glyphs, Selenite pictograms, then some English. I wasn't ready for English words. Maybe it would be repeated.

Although the light danced counterclockwise around the room, the images seemed to linger momentarily, more as a function of the human eye than as physical reality. Furthermore, the symbols appeared from right to left, as in Semitic earth languages. I had noticed that the English words were unreadable. I was ready when they re-appeared.

PLEASE COME IN PEACE

A code? What was Cavor trying to say? Why the mystery?

In a moment I had it. The motion was counterclockwise. The words were backwards.

Good advice.

The light show lasted another fifteen minutes. It revived the Grand Lunar. His eyes were alert, fully open. For him, the images from the beacon were as the sounds of a symphony in the finest theatre on earth. His eyes glistened.

I was silent, waiting his pleasure.

As the images faded, he spoke.

"The human left a package."

A recessed panel on the floor slid open to reveal a jaguar-pelt bag.

"It is time." He closed his eyes. He breathed sharply, and then twitched.

Was he dead?

We left. Sarah refrained from her usual barking.

At the base of the platform, I examined the gift. It was Cavor's notes. The first pages contained the same information as Cavor's broadcasts to earth, as recorded by Wells. During his time with the Selenites, he had great freedom in running experiments and building machines such as the great light show. His last page was a farewell,

> I miss earth. Although I mixed some cavorite for a possible return attempt, conditions have worsened. The Selenites are dying. There is no panic. They are a stoical race. So far we have found no cure. These notes will be placed where they may best be found. No Selenite will ransack the Grand Lunar's eyrie. Everyone knows their duty and is content.

The front flap of Cavor's journal contained directions to his apartment.

His room was small, and contained a mat, a desk and chair, and a few notes on scraps of paper. Everything was tidy, except for a layer of dust. In one corner was a sealed bucket, containing a bluish goo. I dipped in a coin, and held it aloft. Rather than falling to the ground, it drifted to a wall. I caught it.

It was cool and dry. Cavorite worked!

Maybe there is a way home, I hoped, as I lugged the bucket back to Priscilla.

While Sarah roamed, I painted Priscilla's hull with cavorite, letting her dry completely before rolling her over. As I painted, she grew lighter and livelier. I had to tie her down to complete the job.

Without the force of the moon's gravity pulling Priscilla towards the center of the moon, the moon's rotation about her axis should be great enough for the tunnel to behave as a celestial whirligig that would hurl us into space. Air resistance and the remnants of the tunnel's collapsible network, that had cushioned us in landing, would slow us somewhat. Once in space, a launch before nightfall might bring us to within radio reach of earth before our air ran out. I gave no thought to leaving Sarah behind. We were in this together to the end. That was the plan.

For the launch, I attached a long tether to the hatchway. Cutting Priscilla's bonds, I pushed her up the tunnel. She gathered speed. I grabbed the trailing rope and climbed hand over hand, running alongside her, and dove inside. The collapsible network pummeled Priscilla's hull lightly. We kept rising out of the hole. I had pointed the nose of the sub upwards since I preferred seeing where I was going more so than where I had been. Fortunately, cavorite dried in a clear gloss.

Long hours later, we cleared the hole and were free of the moon. Earth was a welcome view. But we weren't rotating. I hadn't anticipated that. We were weightless. Sarah growled. So did I.

As the days passed, earth grew larger. I tried the radio periodically. Three quarters of the way home, I got an answer. It was from the international space station. Earth was tracking me.

"Who are you?"

"I'm the deep water submersible, Priscilla, lost from the recovery ship, Titus, late off the Yucatan."

"You're what? Please say again."

I told them most of the story. I gave them Priscilla's specs, and our status with respect to air, water, and food.

Mission control, after a short period, had a plan.

Within two days, a space ship docked with us. A ship had just happened to be at the station. "How many millions of dollars should be spent on saving one man and his dog," I wondered. Of course, the publicity for the space program would be priceless. I also had some flashy artifacts for the museum: two gold coins, a ceremonial dagger, a complete codex, Verne's canteen, and Cavor's journal, a gateway to understanding a new society. Maybe we could send expeditionary forces down both launchways: down the exit tunnel from earth and down the exit tunnel from the moon. Maybe we could save some Mayans and some Selenites. This was a new day. The long count was beginning afresh.

I opened the hatch.

Sarah barked.

"Welcome home, aquanaut!" greeted the rescuers.

We couldn't save Priscilla. We let her go, off on a tangent to infinity.

Exercises

In the story, objects fall and spin in all the various ways discussed in this book. The exercises below summarize some of this motion. At the end of each exercise is a chapter reference containing ideas and formulas useful in finding the answers.

1. Determine the diameters of the second and third balls on p. 259. Remember that the balls may be neither solid nor of the same material. (Chapter VI)

2. How deep was Priscilla's fall on p. 260? Assume no resistance. (Chapter II or VIII)

3. Imagine that Priscilla's exit speed from earth is 7 mi/sec at the mouth of the launch tunnel and that she underwent a constant acceleration of 10 g. How long is the tunnel and how long did the launch last on p. 266? (Chapter III or IV)

4. Generate some of the coin loops as described by the narrator on p. 267. Recall that Priscilla rotates once every six seconds and has diameter eight feet. (Chapter XII)

5. If Priscilla experiences a constant deceleration in the landing tunnel on the moon, how fast does she ram through the fabric covering the crater on p. 268? Assume that the tunnel proceeds radially from half-way to the moon's center to the moon's surface. (Chapter III)

6. The Mayans are credited with better knowledge of the length of an earth year than as given on p. 258. The long count was first implemented

by the Mayans, not on day zero, but on day 7.13.0.0.0, a day when a 365 day year would have cycled through all the seasons twice since the date 0.0.0.0.0. With this information, calculate how long the Mayans believed an earth year to be. (Chapter XIV)

7. Find the slowest rotation rate for which the beacon in the central strand is not an ordinary point, p. 270. (Chapter XIII)

8. With only the moon's rotation about its axis and with no resistance, how fast will Priscilla exit the moon, p. 272? (Chapter VII)

9. Repeat Exercise 8, factoring in the rotation of the moon about the earth and about the sun. Contrast this value with that of your answer to Exercise 8. (Chapter VII)

10. Using the exit speed of Exercise 9, determine how long it takes Priscilla to rendezvous with the earth, p. 272. (Chapter XI)

preamble xv

O! what a fall was there, my countrymen.
William Shakespeare (1564–1616), *Julius Caesar*

The Big Snowy Mountains are fifty miles west of the University of Wyoming. One spring term when I was a graduate student there, I signed up for a skiing class. Each Wednesday after teaching, I headed for the mountains. It was a welcome break from studies. The air was cold and clean. Sitting in the chair lift, skis dangling, being fifty feet and more above the snow at times, I could see for miles.

My instructor, Ralph, was a graduate student in engineering. He taught us ski safety, techniques of stopping and slowing down, to always be in control, how to gauge the difficulty of a trail, and to be courteous to fellow skiers.

One day I asked about ski-jumping. He raised his eyebrows, "There is an old jump not far from here." We went to a little-used trail where stood a large wooden ramp, showing obvious signs of neglect.

In a few words, Ralph sketched the theory. The ramp is built on fairly level ground just short of a very steep slope. The end of the ramp has an upward slant, causing the speeding jumper to rise, so clearing the remaining level ground, after which the skier falls onto the steep slope, his skis making gentle contact at impact.

Ralph proceeded to demonstrate, and climbed half-way up the ramp. The top half had missing ramp sections. I was uneasy. His planned run looked very short. But Ralph was an expert and an engineer. He knew what he was doing.

"Meet me at the bottom," he said as he took off down the ramp, gaining speed. But not enough. Airborne, he sailed upwards and then came down hard, onto the level ground just before the steep slope. At the impact, his legs, skis, and arms went akimbo, cartwheeling in awful energy. His momentum was enough to carry him over the lip of the veritable cliff. He cascaded in a ball, tumbling hundreds of yards down the slope. The seconds passed like minutes. At the base of the incline, he lay prone in the snow, unmoving.

The slope before me was steeper than any I'd tried before. But there was no one else around to help. I plunged over the precipice, using the skills Ralph had taught, swerving back and forth at the hips, biting into the snow with ski edges perpendicular to the slope. I pulled up, stopping at Ralph's side.

"Are you okay? Are you okay?" I shouted, hoping for the best.

Ralph raised his head slowly, came to his knees, and said, "Yeah, but I think I've done enough teaching for today."

CHAPTER XV

Hesiod's Anvil

Just as Ralph finished teaching for the day, bruised yet breathing and having done his best to show how ski-jumping is done, so too at the end of this book, the time has come to give our answer to the Muses. The Greek poet Hesiod claimed that

> For a brazen anvil falling down from heaven nine nights and days would reach the earth upon the tenth: and again, a brazen anvil falling from earth nine nights and days would reach Tartarus upon the tenth. [**40**, p. 131, lines 722–725]

The little puzzle posed in Chapter I was to devise fall times for the Muses to whisper in Hesiod's ear, ones that would be compatible with our models of the cosmos and the earth.

Hesiod's nine day guess for these falls may sound outlandish. We may feel that heaven is an unbounded distance away from earth or perhaps is in an altogether different realm, and therefore the question is meaningless. Also, the idea of an object falling to the center of the earth from the surface and retaining its identity is pure science fiction.

"How big is the cosmos? How deep is the earth?" Hesiod's answers to these two questions were immense distances, the very idea he wished to convey in his poem. Yet once we quantify a phenomenon, it's natural to wonder if we could do better. History and custom have implicitly set up some conventions we ought to respect when formulating our answer to the Muses.

Convention 1: fall from rest. We shall assume that the initial velocity of the anvil is zero. We also assume that the only mass attracting it is the earth.

Convention 2: conventional nines. We will keep the number nine. The poetry of nine in Ptolemy's nine levels of heaven and Dante's nine levels of Hell cannot be ignored. A fall of nine time units has poetic resonance with

Figure 1. An old woodcut: Hephaestus-Vulcan forging a thunderbolt for Zeus-Jupiter [**82**, entry: incus (anvil)]

history. However we need not stay with days. We could use hours, years, millenia, or any conventional time unit. Thus the fall times of nine units has the satisfying feel of having a unit of time for each of the classic nine levels, both in the heavens and in the earth.

Convention 3: beyond Saturn. We probably need the falling time from heaven to earth to be long enough for the anvil to have been dropped from beyond the known planets of Hesiod's day, somewhere out beyond Saturn. In Ptolemy's cosmos, Saturn's realm lay in the seventh sphere. The stars lay scattered about in the eighth and ninth spheres. Figure 2 in Chapter I depicts a version of Ptolemy's model. Presumably the gates of heaven, the drop site for the anvil, is located somewhere in the ninth sphere.

Convention 4: terrestrial resistance. From Chapter II we saw that it took 19 to 21 minutes for a stone to fall from the earth's surface to its center without resistance, whether we assume a constant gravitational model, a homogeneous earth model, or a geophysical model. There is no readily available conventional nine time units in such a fall time. Therefore, in deference to Convention 2 our anvil must encounter resistance as it falls through the earth.

From heaven to earth

Galileo, the last great physicist of the ancient world and the first of the modern world, discovered that "the distances measured by the falling body increase according to the squares of the time" [35, p. 223]. Galileo extrapolated and calculated that an iron ball falling from rest at the moon's orbit takes 3 hours, 22 minutes, and 4 seconds to travel the supposed 196,000 miles to the earth's center [35, p. 224]. In his thought experiment, Galileo explicitly assumed that gravitational force is independent of the object's distance from the earth, and that all things fall to the earth.

With this assumption, in nine days Hesiod's anvil should fall $D = 2.96 \times 10^9$ km, a distance exceeding that between the sun and Uranus. Since the ancients would have placed Uranus in the eighth sphere had they been able to observe it, the distance D from the earth is probably somewhere in the ninth sphere, a good place for the initial drop point of Hesiod's anvil—the gates of heaven. This calculation might convince us to stay with Hesiod's nine day fall from heaven to earth.

However let's determine the fall time consistent with Newtonian mechanics. Unlike the Galilean model, the Newtonian model assumes that the force of gravity varies inversely with the square of the distance between mass centers. To keep our calculations manageable, we assume that all masses but that of the earth and the anvil are ignored, that heaven's gate is distance D from earth's center, and that the anvil falls in a straight line. Let a, v, t, s be the respective acceleration, velocity, time, and distance of the anvil from the earth's center. Let R and g be the radius of the earth and gravitational acceleration at the surface of the earth, respectively. At $t = 0$, $s = D$ and $v = 0$. As we have done several times before, we start our argument by noting that

$$va = v\frac{dv}{dt} = a\frac{ds}{dt}.$$

Since both v and s are monotonic functions in terms of time while the anvil falls towards the center of the earth,

$$\int_0^t v\frac{dv}{d\tau}\, d\tau = \int_0^t a\frac{ds}{d\tau}\, d\tau$$

or

$$\frac{v^2(s)}{2} = \int_D^s a\, dr,$$

where τ and r are dummy variables for time and distance, respectively. Since $a(s) = -k/s^2$ where $k = gR^2$, the velocity of the anvil in terms of s is

$$v(s) = -\sqrt{\frac{2k(D-s)}{sD}}. \tag{1}$$

Since $v(s)$ is monotonic, the time for the anvil to fall from D to s is

$$t(s) = \int_D^s \frac{1}{v(r)} \, dr.$$

By (1)

$$t(s) = \frac{D^{3/2}}{2\sqrt{2k}} \left(\pi + \frac{2\sqrt{s(D-s)}}{D} - 2\sin^{-1}\sqrt{\frac{s}{D}} \right). \tag{2}$$

Therefore the time T for the anvil to reach the earth is approximately $t(0)$, which from (2) is

$$T = \frac{\pi D^{3/2}}{2\sqrt{2k}} \approx 8970 \text{ years}, \tag{3}$$

where $D = 2.96 \times 10^9$ km, $k = gR^2$, $g = 9.8$ m/sec^2, and $R = 6400$ km. Thus one alternative guess for the Muses is nine thousand years, an answer in accord with all four conventions.

One of the reasons for such a large falling time is that earth's gravitational field was used rather than that of the sun. Adjusting for this greater mass and violating Convention 1, we take $R = 149.6 \times 10^6$ km, the radial distance of the earth from the sun, and $g = 0.00593$ m/sec^2 as the sun's gravitational strength at earth's orbit, and get a modified guess of 15.6 years by (3). If we wish for a more poetic guess of nine years we could move heaven's gate closer to earth so that D is about 2 billion km from the sun, still well beyond Saturn's orbit.

On the other hand, considering the many cinematic productions which have been made over the years in which the Greek gods look down on men and intervene in their lives, we could break Convention 2, and place heaven's gate nine days above the earth's surface. Then (3) gives D as 5.81×10^5 km, about one and a half times the distance between the earth and the moon.

So far we have assumed a straight-line motion model for the anvil. But, according to the ancients, the cosmos rotates about the earth, so presumably heaven's gate rotates about the earth as well. So how does our answer change if the anvil has some tangential velocity? From Chapter IV, the path of an anvil A about a mass centered at the origin O is given by

$$r = \frac{1}{a \cos(\theta + \phi) + \frac{k}{h^2}}, \tag{4}$$

where r is the distance between A and O, θ is the angle between the positive x-axis and A, a is a constant, ϕ is a constant phase angle, and k and h are constants. In particular, k is the gravitational constant as determined by the mass at O. If this mass is the earth, then it induces an acceleration of $f(r) = -k/r^2$ on A when A is r units from O, so if R is the radius of the earth and g is the acceleration due to gravity at the earth's surface, $k = gR^2$. The constant of angular momentum h is $h = r^2 \frac{d\theta}{dt}$, as discussed in Chapter IV. Now if the tangential velocity of the anvil is non-zero initially, its path by (4) is an ellipse that never strikes the earth.

To underscore the difficulties posed by the conventions on the anvil, we revisit Galileo's problem of dropping a ball from the moon. H.G. Wells restated this problem in his 1901 science fantasy, *The First Men in the Moon*, which is analyzed in Chapter XI. In his story, two men build a large hollow glass ball, cover it in a gravity blocking material, and fly off to the moon by following the tangential velocity vector at the earth's launch site. But how can Wells get the spaceship back from the moon? In terms of Hesiod's anvil, what we want to do is release the anvil at the moon's surface and then have the moon disappear, that is, have the moon's gravity disappear. However as we saw in Chapter XI, the moon's rotational rate about its axis is too slow for the anvil to fly along the tangential velocity vector and strike the earth. We can't just drop the anvil from the moon and expect it to reach the earth. We must throw it, and throw it hard. How hard?

Figure 2a shows the anvil's orbit if it is launched at the center of the moon's face to the earth, its tangential initial velocity being given by the moon's rotation about its axis and ignoring all but the earth's gravity. The orbit is almost the same as the moon's orbit. Figure 2b shows the anvil's orbit when it is given an initial tangential velocity of 500 m/sec from the launch site.

For Hesiod's anvil problem, let us drop the anvil at Saturn's orbit, $D = 1.4 \times 10^9$ km from the earth, from a platform traveling at Saturn's orbital period of $T = 29.5$ years about the sun. We can see that the tangential speed (with respect to the platform) with which the anvil is tossed should be about $-2\pi D/T \approx -9.5$ km/sec in order for a highly eccentric elliptical orbit to graze the surface of the earth. Truly, the anvil is well-described as being

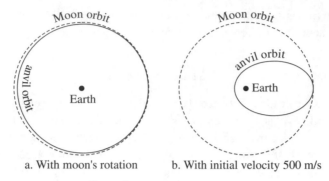

a. With moon's rotation   b. With initial velocity 500 m/s

Figure 2. Orbits of a hurled anvil

"hurled" out of heaven, a phrase used by John Milton to describe Satan's ejection from heaven in *Paradise Lost*.

In summary, nine days, nine thousand years, and nine years all seem to be good answers. The first guess is based upon the constant acceleration model, and the last two are based upon Newton's inverse square model. The first two guesses are based on earth's gravity, and the last one is based upon the sun's gravity. Even though it violates Convention 1, my favorite among these three is nine years.

From earth's surface to its center

An anvil falling through the earth experiences resistance. For the medium of resistance, history offers two choices. Socrates believed that a chasm filled with water ran through the center of the earth [**62**, pp. 381–7]. In the nineteenth century, geophysicists speculated that most of the interior of the earth was molten rock that intermittently emerged as volcanic eruptions. Let us assume that gravity is constant and that a body falling through a uniform medium quickly reaches a terminal velocity. Thus Hesiod's anvil may be dropped down Socrates' watery chasms or down the mouth of some volcano and fall (unmelted) to the earth's center. For the anvil to make the trip in nine days, its terminal velocity must be 6400 km/9 days $\approx$ 8.2 m/sec, a respectable sprinting speed for man.

Now let's see whether this is consistent with Newtonian mechanics. For simplicity, suppose that the anvil A is dropped from the earth's equator. As in Chapter IV, position the polar plane through earth's equator with its center as the origin. With respect to A's position at the polar point (r, θ), the two natural orthogonal unit vectors are $\mathbf{u}_r = (\cos\theta, \sin\theta)$ and $\mathbf{u}_\theta = (-\sin\theta, \cos\theta)$, and A's position is $r\mathbf{u}_r$. A's velocity $\mathbf{v}$ and acceleration $\mathbf{a}$ are

$$\mathbf{v} = \frac{d\,(r, \theta)}{dt} = \frac{d\,r\mathbf{u}_r}{dt} = r\mathbf{u}_\theta \frac{d\theta}{dt} + \mathbf{u}_r \frac{dr}{dt}$$

and

$$\mathbf{a} = \left(r\frac{d^2\theta}{dt^2} + 2\frac{dr}{dt}\frac{d\theta}{dt} \right) \mathbf{u}_\theta + \left(\frac{d^2r}{dt^2} - r\left(\frac{d\theta}{dt}\right)^2 \right) \mathbf{u}_r. \qquad (5)$$

Assume that the only force on the anvil is that induced by the gravitational field of the earth and by resistance due to its velocity. Let $f(r)$ be the gravitational acceleration induced by earth's mass at r units from the origin. Let ω be the rotation rate of the earth. We will assume that the resistance acceleration is proportional to the square of the anvil's speed. Let μ be a positive constant of proportionality. Since $\mathbf{u}_r$ and $\mathbf{u}_\theta$ are orthogonal vectors, the resistance accelerations in each of these unit directions is determined by their respective coefficients of $\mathbf{v}$.

That is, the resistance acceleration in the $\mathbf{u}_r$ direction is proportional to $\left(\frac{dr}{dt}\right)^2$. Since the anvil is falling, r is decreasing, so the resistance acceleration is positive—it tends to hold the anvil up. Thus the resistance acceleration in the $\mathbf{u}_r$ direction is

$$\mu \left(\frac{dr}{dt} \right)^2 \mathbf{u}_r.$$

Since the gravitational acceleration is entirely in the $\mathbf{u}_r$ direction, the total acceleration on the anvil in the $\mathbf{u}_r$ direction is

$$\left(f(r) + \mu \left(\frac{dr}{dt} \right)^2 \right) \mathbf{u}_r. \qquad (6)$$

Let ω be the rotation rate of the earth. The rotation rate of the anvil is $\frac{d\theta}{dt}$, so the relative speed with respect to the rotating earth of the $\mathbf{u}_\theta$ component of $\mathbf{v}$ is $r(\frac{d\theta}{dt} - \omega)$. Thus the resistance acceleration in the $\mathbf{u}_\theta$ direction is

$$-\mu r^2 \left(\frac{d\theta}{dt} - \omega \right)^2 \mathbf{u}_\theta, \qquad (7)$$

negative because the anvil will tend to fall east (counterclockwise) with descent so the resistance acceleration in the $\mathbf{u}_\theta$ direction will be towards the west (clockwise), the negative direction.

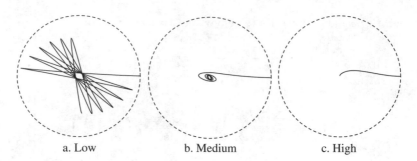

a. Low   b. Medium   c. High

Figure 3. Anvil paths for various resistances

Now equate the $\mathbf{u}_r$ component of (5) with (6), and equate the $\mathbf{u}_\theta$ component of (5) with (7), giving a system of two differential equations governing the motion of the anvil dropped at earth's equator:

$$r\frac{d^2\theta}{dt^2} + 2\frac{dr}{dt}\frac{d\theta}{dt} = -\mu r^2\left(\frac{d\theta}{dt} - \omega\right)^2 \quad \text{and} \quad \frac{d^2r}{dt^2} - r\left(\frac{d\theta}{dt}\right)^2$$

$$= f(r) + \mu\left(\frac{dr}{dt}\right)^2. \tag{8}$$

Since we can't solve the system analytically, let us use the piecewise linear function

$$f(r) = \begin{cases} -\dfrac{g_a s}{a}, & \text{if } 0 \le s < a, \\[2mm] \dfrac{g_a(s-b) - g_b(s-a)}{b-a}, & \text{if } a \le s \le b, \end{cases} \tag{9}$$

with $a = 3500$ km, $b = 6400$ km, $g_a = 10.8$ m/sec², $g_b = 9.8$ m/sec², so a is at the core-mantle boundary and b is at the surface of the earth. This is a fairly accurate representation of the current geophysical understanding of the earth's density.

When we generate graphical solutions to (8), we could plot $r(t)(\cos\theta(t), \sin\theta(t))$, but the resulting paths would be with respect to a nonrotating earth. What we want are paths with respect to the rotating earth,

$$r(t)\big(\cos(\theta(t) - \omega t), \sin(\theta(t) - \omega t)\big),$$

where ω is the rotational speed of the earth, 2π radians per day.

Plotting solutions for various μ values yields Figure 3, where the dashed circles represent the earth's circumference. In a 1679 exchange of letters, Isaac Newton and Robert Hooke, discussing the problem of a ball

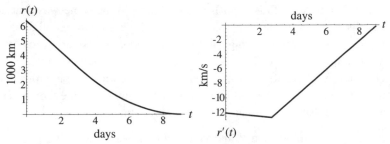

a. Position $r(t)$ with respect to time   b. Falling speed $r'(t)$ with respect to time

Figure 4. The falling body

falling down a hole through the earth and experiencing resistance, essentially sketched Figures 3b and 3a [6, pp. 18–20]. There are two solutions for which the anvil reaches the earth's center in nine days, one corresponding to a very low resistance and the other to a high resistance.

The high value of $\mu \approx 67$/km gives a fall time of nine days. The graphs of $r(t)$ and $r'(t)$ are in Figure 4. As time goes on, the downward speed of the anvil increases until 2.7 days, when it achieves a maximum velocity of about -12.6 m/sec at the core-mantle juncture, after which it decreases linearly to 0 on day 9.

With these assumptions, Newtonian mechanics and Hesiod's Muses concur. A fall time of nine days through the earth is a reasonable guess for the anvil to reach earth's center. See Exercise 5 for an exploration of a fall time of nine hours.

Some parting observations

Our advice for the Muses is to change the heaven-to-earth fall time of the anvil to nine years and to keep the earth-to-its-center (or earth to Tartarus) fall time of the anvil as nine days. Hesiod probably would be troubled by this suggestion, since the clash of units destroys poetic symmetry. I like to think that even if the Muses followed my advice, Hesiod would rub his ears, scratch his head, and persuade himself that he really had heard nine days for each fall. "It's only a poem after all," he might say.

However, despite being merely the lines in a poem, within Hesiod's twin claims about falling times is a significant idea far ahead of his time, that the mechanism governing falling through the heavens is the same as that governing falling near the earth's surface. Though commonplace today, it would have been revolutionary from Hesiod's time until Galileo's. For, from thinking that an anvil falls from heaven's zenith in the same way that an anvil falls when dropped down a chasm on earth, it is a small step to think of the planets themselves as falling. And once we think that, we are ready for a Newton to unveil the calculus.

Of course, shaking off a key traditional notion and replacing it with a new notion is almost always tumultuous. Voltaire summarized the Copernicus Galileo Kepler Newton ultimate triumph of ideas this way:

> A blushing Dominican at Rome said to an English philosopher [New-ton], "You are a dog; you say it is the earth that turns round, never reflecting that Joshua made the sun stand still." "Well! my reverend father," replied the other, "and since that time the sun has been immov-able." The dog and the Dominican embraced each other; and even the Italians [the Church] were, at last, convinced that the earth turns round [**92**, pp. 201–202].

In understanding the championing of one idea over another, it is good to remember that the old Ptolemaic system, despite its flaws, accurately pre-dicted planetary motion, and that, prior to Newton, the prevailing model of the universe can best be described as Socratic: that reality—or the laws of motion—changes according to a hierarchy in the heavens; that "things in the upper world [are] more superior to those in this world of ours," [**62**, p. 377]. After Einstein, the prevailing model has returned to this old Socratic model. For example, the motion of very small (sub-atomic) objects is qualitatively different than the motion of thrown stones or falling planets. And at rela-tivistic speeds—speeds near that of light—time and mass dilate in a strange way. Today's physicists are trying to find a unified theory of motion, as the current multi-modeled one seems too ungainly to be the best of all possible perspectives.

Nevertheless, Hesiod's idea that the nature of motion is invariant from heaven above to the underworld below has blossomed into analytical formu-las whose approximations to reality are remarkably accurate for a wide range of falling, spinning motion—an achievement which this book both celebrates and explores.

Exercises

1. As paraphrased by Archimedes, Aristarchus proposed that

> the sphere of the fixed stars, situated about the same centre as the sun, is so great that the circle in which he [Aristarchus] supposes

the earth to revolve has such a proportion to the distance of the fixed stars as the centre of the sphere bears to its surface [**86**, pp. 3–5].

Interpret the phrase "the centre of the sphere bears to its surface" as "the radius of the sun is to the radial distance of the earth from the sun." Under this assumption, use the rule of Aristarchus to find the radius of the sphere of the fixed stars. (Use 149.6×10^6 km as the distance of the earth from the sun and 6.96×10^5 km as the radius of the sun.)

2. Under the assumption that acceleration is constantly 9.8 m/sec^2 towards the sun, determine how long it takes an anvil to fall from the surface of the sphere of fixed stars to the sun, using the hypothesis of Aristarchus in Exercise 1.

3. Derive equation (3) directly from Kepler's third law of planetary motion.

4. Generate the graph corresponding to Figure 3 using the standard acceleration function $f(r) = -gr/R$ rather than (9), where $g = 9.8$ m/sec^2.

5. In Dante's *Inferno*, Dante and Virgil travel from the earth's center to the surface (at the Antipodes) in 9.5 hours. Evidently they are not delayed by conversation as on their descent. Use the falling rate of 4000 miles in 9.5 hours to generate a graph corresponding to Figure 4a.

6. Suppose that the resistance component of (6) varies according to density $\delta(r)$ as well as the square of the descent rate, namely, according to $\mu\delta(r)(\frac{dr}{dt})^2$, and where the graph of $\delta(r)$ is a straight line. Assume that the density of the earth is 5000 kg/m^3 at the earth's surface and is 13,000 kg/m^3 at the earth's center. Determine μ so that the anvil descends to earth's center in nine days.

7. Repeat Exercise 6 with the density function as defined in the graph of Figure 9 in Chapter II.

8. In *Alice in Wonderland*, Alice falls down a rabbit hole for a long time— let us say 30 minutes. Suppose that Wonderland is half-way to the earth's center. From this information, determine μ of (6).

9. In Milton's *Paradise Lost*, Satan is "hurled headlong flaming from th' ethereal sky" in a fall lasting nine days. He evidently encounters resistance during his fall. This event occured before earth was formed. Imagine that the solar system was a uniform, spherical dust cloud with (unknown) radius R km and mass 2×10^{30} kg. Suppose that Satan, at the

boundary of this dust cloud, is given an initial velocity of 1000 km/sec towards the center C of this cloud. Determine possible values of R and μ so that Satan reaches C in nine days.

10. The space probe Galileo of Figure 5, launched from the shuttle Atlantis in 1989, performed its mission of collecting data until it plunged into Jupiter's atmosphere in September 2003. Ignoring incineration and Jovian turbulence, the probe followed a path modeled by (8), as do any meteorites falling through Jupiter's atmosphere, until they burn up or hit something solid. Imagine that Galileo is about to plunge into

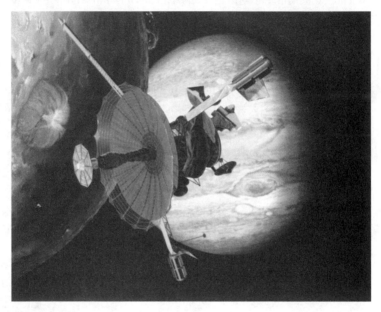

Figure 5. Galileo probe prior to plunging into Jupiter (courtesy NASA/JPL-Caltech)

Jupiter. At time $t = 0$, its motion is in Jupiter's equatorial plane: its position is $r(0) = R = 68700$ km, $\theta(0) = 0$ and its velocity is $r'(0) = -0.1$ km/sec, $\theta'(0) = 1.1\Omega$, where Jupiter's rotation about its axis is $\Omega = 2\pi$ radians/10 hours. The acceleration due to gravity at radius R is 2.74 times as great as earth's surface gravity. How long will it take for Galileo to penetrate 1000 km nearer to Jupiter's center?

Appendix

The following material comes from linear algebra, vector calculus, and differential equations. For greater detail, consult any standard text.

1. Natural directions

In $\mathcal{R}^3$ the three natural unit directions are denoted $\mathbf{i} = (1, 0, 0)$, $\mathbf{j} = (0, 1, 0)$, and $\mathbf{k} = (0, 0, 1)$, as in Figure 1. These vectors point in the positive x, y, and z directions.

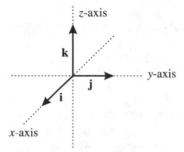

Figure 1. The natural directions

2. Columns, rows, transposes, and vector notation

A three-dimensional vector $\mathbf{u}$ in $\mathcal{R}^3$ is generally meant to be interpreted as a column of three real-number components, u_1, u_2, and u_3,

$$\mathbf{u} = \begin{bmatrix} u_1 \\ u_2 \\ u_3 \end{bmatrix},$$

although we sometimes abuse the convention for convenience and write $\mathbf{u} = (u_1, u_2, u_3)$ when confusion cannot arise.

The transpose of a column vector $\mathbf{u}$, denoted $\mathbf{u}^\mathsf{T}$, is a row vector and vice versa:

$$\mathbf{u} = \begin{bmatrix} u_1 \\ u_2 \\ u_3 \end{bmatrix} = (u_1, u_2, u_3)^\mathsf{T} \quad \text{and} \quad \mathbf{u}^\mathsf{T} = \begin{bmatrix} u_1 \\ u_2 \\ u_3 \end{bmatrix}^\mathsf{T} = (u_1, u_2, u_3).$$

In the text, to label vectors (directions, velocities, and accelerations) we use bold face letters, while to label positions (the coordinates of points) we use ordinary face letters. Sometimes we abuse this distinction, and refer to the vector u when u is a position, and refer to the position $\mathbf{v}$ when $\mathbf{v}$ is a vector. When we use this convention, we mean the vector whose head is u and whose tail is the origin and the position whose coordinates are the components of $\mathbf{v}$.

3. Linear independence

A set of n vectors $\{\mathbf{u}_i\}_{i=1}^n$ is a linearly independent set if whenever

$$\sum_{i=1}^n \lambda_i \mathbf{u}_i = \mathbf{0}, \ \lambda_i = 0 \text{ for all } i, \ 1 \le i \le n.$$

This definition is the fundamental notion of linear algebra. The entire subject could be looked at as a study of writing this phrase in different ways. In $n = 2$, $\mathbf{u}_1$ and $\mathbf{u}_2$ are linearly independent if $\mathbf{u}_1$ is not a real-numbered multiple of $\mathbf{u}_2$. Similarly, we say that two non-zero functions are linearly independent if one is not a multiple of the other.

4. Dot product

Let $\mathbf{u} = (u_1, u_2, u_3)$ and $\mathbf{v} = (v_1, v_2, v_3)$ be vectors in $\mathcal{R}^3$. The dot product of $\mathbf{u}$ and $\mathbf{v}$, denoted $\mathbf{u} \cdot \mathbf{v}$, is

$$\mathbf{u} \cdot \mathbf{v} = u_1 v_1 + u_2 v_2 + u_3 v_3. \tag{10}$$

For example, the dot product of $\mathbf{p} = (1, 2, 3)$ and $\mathbf{q} = (2, 3, -1)$ is 5. The dot product can be used to find the length or magnitude of a vector. The magnitude of $\mathbf{u}$, denoted $||\mathbf{u}||$, is

$$||\mathbf{u}|| = \sqrt{\mathbf{u} \cdot \mathbf{u}} = \sqrt{u_1^2 + u_2^2 + u_3^2}. \tag{11}$$

Another way to write (11) is $||\mathbf{u}||^2 = \mathbf{u} \cdot \mathbf{u}$. A unit vector in the $\mathbf{u}$ direction is $\mathbf{u}/||\mathbf{u}||$. Let θ be the angle between vectors $\mathbf{u}$ and $\mathbf{v}$, as shown in Figure 2. The dot product can be used to determine θ:

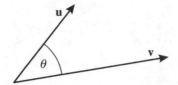

Figure 2. The angle between two vectors

$$\cos\theta = \frac{\mathbf{u} \cdot \mathbf{v}}{||\mathbf{u}|| \, ||\mathbf{v}||}. \tag{12}$$

For example, the angle between $\mathbf{A}$ and $\mathbf{B}$ by (12) is $\cos^{-1}(5/14) \approx 1.2056$ radians, or about $69.08°$. Two vectors are perpendicular to each other if $\theta = \pi/2$, so from (12)

$$\mathbf{u} \perp \mathbf{v} \text{ if and only if } \mathbf{u} \cdot \mathbf{v} = 0. \tag{13}$$

We say that two vectors $\mathbf{u}_1$ and $\mathbf{u}_2$ are orthogonal if $\mathbf{u} \perp \mathbf{v}$, and are orthonormal if they have unit magnitudes as well as being orthogonal.

5. Projections

Oftentimes we wish to project a vector $\mathbf{u}$ onto a unit vector $\mathbf{w}$, as shown in Figure 3. The projection $\mathbf{p}$ of $\mathbf{u}$ onto $\mathbf{w}$ is the vector in the $\mathbf{w}$ direction whose

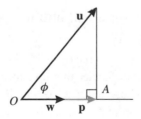

Figure 3. Projecting $\mathbf{u}$ onto $\mathbf{w}$

tail is the common tail of $\mathbf{u}$ and $\mathbf{w}$ and whose head is at A. If ϕ is the angle between $\mathbf{u}$ and $\mathbf{w}$, then (12) gives $\cos\phi = \mathbf{u} \cdot \mathbf{w}/||\mathbf{u}||$, since $\mathbf{w}$ is a unit vector. The length p of the projection is thus

$$p = ||\mathbf{u}|| \cos\phi = \mathbf{u} \cdot \mathbf{v}. \tag{14}$$

When p is negative, the projection of $\mathbf{u}$ onto $\mathbf{w}$ lies in the $-\mathbf{w}$ direction. Thus, the projection of $\mathbf{u}$ onto $\mathbf{w}$ is $\mathbf{p} = (\mathbf{u} \cdot \mathbf{w})\mathbf{w}$.

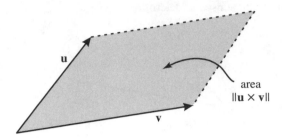

Figure 4. Area of a parallelogram

6. Cross product

The cross product of the three-dimensional vectors $\mathbf{u}$ and $\mathbf{v}$ is another three-dimensional vector, denoted $\mathbf{u} \times \mathbf{v}$, given by

$$\mathbf{u} \times \mathbf{v} = (u_2 v_3 - u_3 v_2)\mathbf{i} - (u_1 v_3 - u_3 v_1)\mathbf{j} + (u_1 v_2 - u_2 v_2)\mathbf{k}. \qquad (15)$$

It is perpendicular to both $\mathbf{u}$ and $\mathbf{v}$:

$$\mathbf{u} \cdot (\mathbf{u} \times \mathbf{v}) = 0 = \mathbf{v} \cdot (\mathbf{u} \times \mathbf{v}).$$

For example the cross product of $\mathbf{u} = (1, 2, 3)$ and $\mathbf{v} = (2, 3, -1)$ is $\mathbf{u} \times \mathbf{v} = (-11, 7, -1)$, and $\mathbf{u} \cdot (\mathbf{u} \times \mathbf{v}) = \mathbf{v} \cdot (\mathbf{u} \times \mathbf{v})$. The area of the parallelogram determined by $\mathbf{u}$ and $\mathbf{v}$ is $\|\mathbf{u} \times \mathbf{v}\|$.

7. Area element on a spherical bubble

Consider the mapping $\mathbf{r} : \mathcal{R}^2 \rightarrow \mathcal{R}^3$, given by the spherical coordinate transformation, $\mathbf{r}(\theta, \phi) = r(\cos\theta \sin\phi, \sin\theta \sin\phi, \cos\phi)$ where r is a fixed positive real number. Under this transformation, a small rectangular patch in the plane, whose sides have length $\Delta\theta$ and $\Delta\phi$ and whose sides are parallel to the θ and ϕ axes, is transformed into a patch on the sphere of radius r that is approximately a parallelogram with sides of length $\frac{\partial \mathbf{r}}{\partial \theta}\Delta\theta$ and $\frac{\partial \mathbf{r}}{\partial \phi}\Delta\phi$, where $\frac{\partial \mathbf{r}}{\partial \theta} = r(-\sin\theta \sin\phi, \cos\theta \sin\phi, 0)$ and $\frac{\partial \mathbf{r}}{\partial \phi} = r(\cos\theta \cos\phi, \sin\theta \cos\phi, -\sin\phi)$, as sketched in Figure 5. By (15), the area of this patch is approximately

$$\left\| \frac{\partial \mathbf{r}}{\partial \theta} \times \frac{\partial \mathbf{r}}{\partial \phi} \right\| \Delta\theta \, \Delta\phi,$$

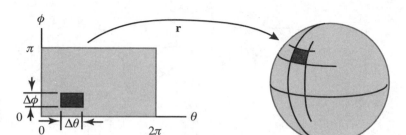

Figure 5. The spherical transformation

which, by (15) and (11), is $r^2 \sin\phi \, \Delta\theta \, \Delta\phi$. This is the area element dA when using $\mathbf{r}$ to integrate on the surface of a sphere. The transformation $\mathbf{r}$ changes area, from its domain to its range, by the factor $r^2 \sin\phi$. When we use $\mathbf{r}$ to integrate a real-valued function f over the surface of a sphere, we integrate in the plane by going backward through $\mathbf{r}$. Thus, before integrating, we must multiply f by this factor in order to nullify the effect of the functional inverse of $\mathbf{r}$. In particular, if S represents the spherical shell of Figure 5, then

$$\iint\limits_{S} f(x, y, z) \, dA = \int_0^\pi \int_0^{2\pi} f(\mathbf{r}(\theta, \phi)) r^2 \sin\phi \, d\theta d\phi.$$

Area element on a cylindrical shell Consider the mapping $\mathbf{r} : \mathcal{R}^2 \to \mathcal{R}^3$, given by the cylindrical coordinate transformation, $\mathbf{r}(\theta, z) = (r\cos\theta, r\sin\theta, z)$ where r is a fixed positive real number. Under this transformation, a small rectangular patch in the plane, whose sides have length $\Delta\theta$ and Δz and whose sides are parallel to the θ and z axes, is transformed into a patch on the cylinder of radius r that is approximately a parallelogram with sides of length $\frac{\partial \mathbf{r}}{\partial\theta} \Delta\theta$ and $\frac{\partial \mathbf{r}}{\partial z} \Delta z$, where $\frac{\partial \mathbf{r}}{\partial\theta} = (-r\sin\theta, r\cos\theta, 0)$ and $\frac{\partial \mathbf{r}}{\partial z} = (0, 0, 1)$, as sketched in Figure 6, (where h is the height of the cylinder). By (15), the area of this patch is approximately

$$\left\| \frac{\partial \mathbf{r}}{\partial\theta} \times \frac{\partial \mathbf{r}}{\partial z} \right\| \Delta\theta \, \Delta z,$$

which, by (15) and (11), is $r \, \Delta\theta \, \Delta z$. This is the area element dA when using $\mathbf{r}$ to integrate on the surface of a cylinder. If C represents the cylindrical shell of Figure 6, then

$$\iint\limits_{C} f(x, y, z) \, dA = \int_0^h \int_0^{2\pi} f(\mathbf{r}(\theta, z)) r \, d\theta dz.$$

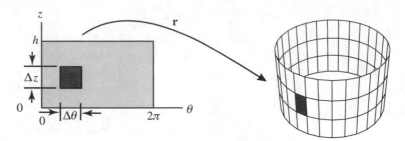

Figure 6. The cylindrical transformation

Area element on a disk Let h and R be positive real numbers and let
s be the mapping $\mathbf{s} : \mathcal{R}^2 \rightarrow \mathcal{R}^3$, given by $\mathbf{s}(\theta, r) = (r \cos \theta, r \sin \theta, h)$.
Under this transformation, a small rectangular patch in the plane, whose
sides have length $\Delta \theta$ and Δr and whose sides are parallel to the θ and r axes,
is transformed into a patch on the top of a cylinder of height h and radius R
that is approximately a parallelogram with sides of length $\frac{\partial \mathbf{s}}{\partial \theta} \Delta \theta$ and $\frac{\partial \mathbf{s}}{\partial r} \Delta r$,
where $\frac{\partial \mathbf{s}}{\partial \theta} = (-r \sin \theta, r \cos \theta, 0)$ and $\frac{\partial \mathbf{r}}{\partial r} = (\cos \theta, \sin \theta, 0)$, as sketched in
Figure 7.

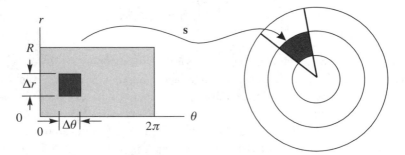

Figure 7. The cylindrical transformation

By (15), the area of this patch is approximately

$$\left\| \frac{\partial \mathbf{s}}{\partial \theta} \times \frac{\partial \mathbf{s}}{\partial r} \right\| \Delta \theta \, \Delta r,$$

which, by (15) and (11), is $r \, \Delta \theta \, \Delta r$. This is the area element dA when using **s**
to integrate on the top of the cylinder. If T represents the top of this cylinder
as in Figure 7, then

$$\iint_T f(x, y, z) \, dA = \int_0^R \int_0^{2\pi} f(\mathbf{r}(\theta, z)) r \, d\theta dr.$$

8. Simple Harmonic Motion

We say that a differential equation is homogeneous if the zero function is a solution of the differential equation. The differential equation modeling simple harmonic motion is

$$y'' + k^2 y = 0, \tag{16}$$

where k is a positive number. We solve (16) by following Brenner [**14**, p. 344]. Let $t = kx$ so that the equation becomes $y'' + y = 0$ where the differentiation is with respect to t rather than x. Let

$$A = y \cos t - y' \sin t \quad \text{and} \quad B = y \sin t + y' \cos t.$$

If we differentiate both equations and use $y'' + y = 0$, we get $A' = 0 = B'$, so A and B are constants. Then eliminate y' from the pair of equations giving the solution to (16) as

$$y = A \cos(kx) + B \sin(kx). \tag{17}$$

In general, for any second-order homogeneous differential equation E, E can be written as a linear combination of two linearly independent functions. Since $\sin(kx)$ is not a multiple of $\cos(kx)$, (17) is the general solution to (16).

Using the identity $\cos(\theta + \omega) = \cos \theta \cos \omega - \sin \theta \sin \omega$, we can write (17) as

$$y = a \cos(kx + \phi), \tag{18}$$

where ϕ is a phase angle and a is a constant. To solve the equation

$$y'' + k^2 y = C, \tag{19}$$

we use the result that the general solution to any non-homogeneous differential equation is the sum of the general solution to the corresponding homogeneous solution and a particular solution to the non-homogeneous differential equation. Since $y = C/k^2$ is a particular solution to (19), the general solution to (19) is

$$y = a \cos(kx + \phi) + \frac{C}{k^2}.$$

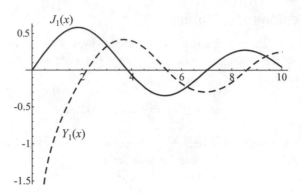

Figure 8. The Bessel functions

9. Bessel functions

The two linearly independent solutions to the Bessel differential equation

$$x^2 y'' + xy' + (x^2 - 1)y = 0$$

are $J_1(x)$ and $Y_1(x)$ whose series representations are

$$J_1(x) = \sum_{n=0}^{\infty} \frac{(-1)^n x^{2n+1}}{2^{2n+1} n!(n+1)!}$$

and

$$Y_1(x) = \frac{2}{\pi} \left(\gamma + \ln \frac{x}{2} \right) J_1(x) - \frac{2}{\pi x} - \frac{1}{\pi} \sum_{n=0}^{\infty} \frac{(-1)^n (H_n + H_{n+1})}{n!(n+1)!} \left(\frac{x}{2} \right)^{2n+1},$$

where

$$H_n = 1 + \frac{1}{2} + \frac{1}{3} + \cdots + \frac{1}{n}$$

and

$$\gamma = \lim_{n \to \infty} (H_n - \ln n) \approx 0.57722$$

is Euler's constant. Most computer algebra systems have Bessel functions built in, just as sines and cosines are. Figure 8 shows that they look like damped sine and cosine curves.

10. Matrices

The transpose of a 3×3 matrix A is another 3×3 matrix, denoted A^T:

$$A^T = \begin{bmatrix} a_{11} & a_{12} & a_{13} \\ a_{21} & a_{22} & a_{23} \\ a_{31} & a_{32} & a_{33} \end{bmatrix}^T = \begin{bmatrix} a_{11} & a_{21} & a_{31} \\ a_{12} & a_{22} & a_{32} \\ a_{13} & a_{23} & a_{33} \end{bmatrix}.$$

The rows of A are the columns of A^T.

The product of a 3×3 matrix A and a 3-dimensional column vector $\mathbf{u}$ is

$$A\mathbf{u} = \begin{bmatrix} a_{11} & a_{12} & a_{13} \\ a_{21} & a_{22} & a_{23} \\ a_{31} & a_{32} & a_{33} \end{bmatrix} \begin{bmatrix} u_1 \\ u_2 \\ u_3 \end{bmatrix} = \begin{bmatrix} a_{11}u_1 + a_{12}u_2 + a_{13}u_3 \\ a_{21}u_1 + a_{22}u_2 + a_{23}u_3 \\ a_{31}u_1 + a_{32}u_2 + a_{33}u_3 \end{bmatrix},$$

a 3-dimensional column vector whose components are the rows of A dotted with $\mathbf{u}$.

The inverse of a 3×3 matrix A, if it exists, is denoted by A^{-1}, and is that 3×3 matrix for which $AA^{-1} = I = A^{-1}A$, where I is the identity matrix,

$$I = \begin{bmatrix} 1 & 0 & 0 \\ 0 & 1 & 0 \\ 0 & 0 & 1 \end{bmatrix}.$$

A 3×3 matrix A is an isometry if $\|A\mathbf{u}\| = \mathbf{u}$ for all vectors $\mathbf{u}$ in $\mathcal{R}^3$. The only isometries that correspond to transformations of the underlying vector space by matrix multiplication are rotations about the origin or reflections about a line through the origin. The matrix multiplication corresponding to a counterclockwise rotation of the plane about the origin by an angle θ is

$$M = \begin{bmatrix} \cos\theta & -\sin\theta \\ \sin\theta & \cos\theta \end{bmatrix}.$$

In Figure 9 suppose that the point $P = (x, y)$ is rotated counterclockwise by an angle of θ to the point Q. We wish to find the coordinates of Q in

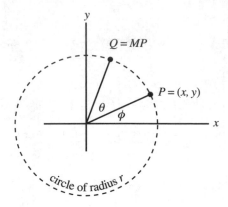

Figure 9. A counterclockwise rotation by θ

terms of r, x, y, θ, and ϕ. From the figure, $x = r \cos \theta$ and $y = r \sin \theta$ so $Q = (r \cos(\phi + \cos), r \sin(\phi + \theta))$. By the trigonometric addition formulas,

$$Q = \begin{bmatrix} r \cos \phi \cos \theta - r \sin \phi \sin \theta \\ r \sin \phi \cos \theta + r \cos \phi \sin \theta \end{bmatrix}$$

$$= \begin{bmatrix} \cos \theta\, x - \sin \theta\, y \\ \cos \theta\, y + \sin \theta\, x \end{bmatrix} = \begin{bmatrix} \cos \theta & -\sin \theta \\ \sin \theta & \cos \theta \end{bmatrix} \begin{bmatrix} x \\ y \end{bmatrix}.$$

That is, $Q = MP$. The matrix

$$N = \begin{bmatrix} \cos \theta & -\sin \theta & 0 \\ \sin \theta & \cos \theta & 0 \\ 0 & 0 & 1 \end{bmatrix}$$

is the matrix for which $N\mathbf{u}$ is a rotation of $\mathbf{u}$ counterclockwise about the positive z-axis by θ radians, because $N\mathbf{u}$ changes the first two components by a multiplication of M and leaves the third component of $\mathbf{u}$ unchanged.

Cast of Characters

Alice is a character created in 1865 by Charles Dodgson (pen name: Lewis Carroll) who modeled her after Alice Liddell, the young daughter of a good friend. Dodgson (1832–1898) was a mathematician who delighted in paradoxes and puzzles; when he had collected enough for a book, he set them all into a story.

Archimedes (287–212 BC), the greatest of the ancient mathematicians, discovered specific gravity while in a bathtub, and then ran naked through the streets shouting, "Eureka!" Among his achievements, he demonstrated that a mass of liquid water will assume a spherical shape, given that all things fall toward a specific point.

Aristotle (384–322 BC) emphasized that observation is a key to understanding nature. Along with Plato, their writings form the core of ancient western philosophy.

Roger Bacon (1214–1294) was a professor and monk who championed Artistotle's empirical method of understanding the universe through experimentation. Towards the end of his life, though forbidden by his Franciscan order to write, Bacon was encouraged by the pope to proceed in secret. Bacon did so, producing his major work—*Opus Majus*—which was promptly condemned under the next pope.

Isaac Barrow (1630–1677) was one of Newton's teachers, and was the first of the Lucasian professors at Cambridge University. Newton followed him as the second holder of this distinguished chair.

Friedrich Bessel (1784–1846) was a German mathematician best known for determining distances to nearby stars using the method of parallax, a method based on star positions taken at various times during earth's orbit about the sun. In his work, he used what is now known as Bessel differential equations, solutions to which were discovered earlier by Daniel Bernoulli (1700–1782).

Irl Bivens (b. 1951) is a master juggler of four balls, and is working on adding a fifth. He discovered the Bivens parametrization of Chapter XIII.

Tycho Brahe (1546–1601) was a Danish astronomer who melded together the Ptolemaic system and the Copernican system so as to dovetail with motion data of the planets. His most famous student was Kepler. At age twenty, Brahe lost most of his nose in a duel, and so wore artificial ones made of gold, silver, or copper, as social occasions warranted.

Edgar Rice Burroughs (1875–1950) is one of the few writers who made a fortune writing for the pulp science fiction media. His most famous character is Tarzan.

Arthur C. Clarke (b. 1917) is best known for the 1968 movie *2001* featuring rotating space stations and an artificial intelligence named HAL. He is credited with the idea in 1945 for geostationary communication satellites.

Confucius (551–479 BC) was a political advisor to various feudal state rulers in China, advocating social justice. After being jaded by his lack of success, he became an itinerant teacher, and attracted a considerable following among the younger generation.

Nicolaus Copernicus (1473–1543) privately distributed a heliocentric theory to his friends in 1514, and subsequently consented to a public printing. Legend has it that on his death bed he was presented with a just-off-the-press copy of his great work, *De revolutionibus*, whereupon he awoke from a coma before passing on.

Dante Alighieri (1265–1321) was a politician who, when exiled from his beloved Florence, became a poet and produced the most influential book of the Middle Ages, *The Divine Comedy*.

Leonardo da Vinci (1452–1519) was the quintessential renaissance man, from sculpture, to paintings, to anatomy, to designing war machines. In his youth he was apprenticed to an artisan and was never formally instructed in Latin or mathematics. He was left-handed, and used a mirror to enable himself to write smoothly.

René Descartes (1596–1650) popularized the coordinate system of the xy-plane. He also championed the idea that no force could cross a vacuum, which in turn meant that the universe was filled with matter, an idea that lasted fifty years after Newton's *Principia*.

Arthur Conan Doyle (1859–1930) was a medical doctor before gaining fame as a writer, most notably by creating the Sherlock Holmes character.

Leonhard Euler (1707–1783) was the most prolific of all mathematicians. His work fills 75 volumes, half of which he generated after he became blind.

Pierre Fermat (1601–1665) was a French statesman who enjoyed mathematical correspondence throughout his career. His first letters involved a critique of Galileo's notion of free fall.

Galileo Galilei (1564–1642) was an experimenter who published his findings in Italian. He championed the heliocentric system of Copernicus. In 1610, through his telescope, he discovered four moons rotating about Jupiter.

George Gamow (1904–1968) was a Russian physicist interested in the origin of the universe. Together with Ralph Alpher, he published a paper on cosmogenesis. To make a pun, they added a fictitious name to their author list, Hans Bethe, so that their theory was referred to as the Alpher-Bethe-Gamow theory, or the α-β-γ theory.

Edmond Halley (1656–1742), scientist extraordinaire, convinced Newton to publish the *Principia*. Among his achievements, he speculated that the comets of 1456, 1531, 1607 and 1682 were one and the same and predicted its return in 1758. Duly arriving in 1758, it was thereafter called Halley's comet.

Stephen Hawking (b. 1942) is the current Lucasian professor of mathematics at Cambridge University. He discovered black holes. Despite being severely handicapped by motor neurone disease since young adulthood, he is a best-selling author of popular mathematics and physics.

Hesiod (circa 700 BC) wrote the *Theogeny* which catalogues the entire pantheon of Greek gods. Tradition has it that Hesiod and Homer competed in poetry contests.

Homer (circa 750 BC) was a legendary, itinerant, blind poet who is credited with composing the *Odyssey* and the *Iliad*.

Robert Hooke (1635–1703) was the greatest experimental scientist of the seventeenth century. As curator to the Royal Society for many years, his job was to demonstrate a new experiment at each regular meeting.

Christiaan Huygens (1629–1695) was an enterprising Dutch mathematician. Using a telescope, he was the first to observe the Orion Nebula, and he proposed that Saturn's rings consisted of rocks. Besides building the first pendulum clock, he invented the pocket watch.

Johannes Kepler (1571–1630) was a student of Tycho Brahe. From Brahe's data, Kepler surmised that planets orbit the sun in elliptical paths. He was the first to conjecture that the moon caused the tides on earth. In the *Somnium*, describing a trip to the moon, Kepler wrote footnote commentaries that are longer than the story itself.

Stanislaw Lem (b. 1921), a Polish satirist, used science fiction as a means to critique the Cold War, a maneuver that outwitted the censors of his day and allowed him great latitude in social commentary. After all, it's hard for a regime to feel insulted directly by the injustice, ineptitude, and sterility within a fictional society far, far away in time and space even if it is a mirror of itself.

John Milton (1608–1674) was an English poet, best known for *Paradise Lost*, written after he had become blind. This blindness also spared his life, for after the demise of Cromwell's Revolution, many intellectuals who had supported the English republic were summarily executed. Evidently the powers that be decided that blindness was equivalent to death.

Isaac Newton (1643–1727), along with Gottfried Wilhelm von Leibniz (1646–1716), is credited with founding calculus. Newton's *Principia* was published in 1687, being a derivation of Kepler's laws under the assumption of an inverse square law for gravity.

Plato (427–347 BC) emphasized that knowledge comes from within, from dialogue, and that observation is merely shadowy glimpses of reality. He was a disciple of Socrates. Along with his most famous student Aristotle, their writings form the core of ancient western philosophy.

Plutarch (46–127), a Greek historian and essayist, is most famous for contrasting Greek and Roman lives of noble character. In an essay on the moon, he speculates about creatures roaming around on its surface.

Edgar Allan Poe (1809–1849) was a master of the short story, verse, and literary criticism, and is considered to be an early pioneer of science fiction, detection, and horror genres.

Terry Pratchett (b. 1948) is the creator of Discworld, as illustrated in Figure 1 in Chapter I, in which he parodies society with great commercial success.

Ptolemy (100–178), also known as Claudius Ptolemaeus, codified the Greek view of the cosmos in the *Almagest* as a series of spheres about the earth, as shown in Figure 2 in Chapter I.

Rudy Rucker (b. 1946), well known for his cyper-punk style science fiction, began his writing career by popularizing mathematics. His novel, *Spaceland*, (2002), is a romance in the fourth dimension, being a new twist on Edwin Abbott's classic *Flatland*, (1884), a romance of life in two dimensions.

Satan and **Lucifer** are different names for a fallen archangel.

Willebrord Snell (1580–1626) was a Dutch mathematician who discovered the law of light refraction. He also found a way to determine distances between two points on the earth using triangularization and latitude measurements.

Socrates (470–399 BC) was convicted of impiety and corrupting Athenian youth, and chose death rather than ostracism. His life-long habit was continual conversation, to question everything. Much of what we know about Socrates comes through his most famous student, Plato.

Jules Verne (1828–1905) was the first to make a fortune writing science fiction romances. He predicted air travel, space travel, submarines, cars, and instant communication long before they were realized.

Virgil (70–19 BC) was a master poet of Rome. His most famous epic is the *Æneid*. Dante admired his poetry so much that he made him his figurative guide in the *Inferno*.

Voltaire (1694–1778) was the pen name of François-Marie Arouet, a French wit and man of letters. He is widely considered to be the most influential intellect of his time.

H. G. Wells (1866–1946) wrote scientific romances in his early years, some of which were sparked by reading Verne's stories. In later years, he was a champion of socialist utopias.

Eugene Wigner (1902–1995) won the Nobel Prize in 1963 for his work on determining the subatomic structure of matter. He popularized the phrase, the unreasonable effectiveness of mathematics.

Winnie the Pooh is an animated stuffed bear character created by novelist A. A. Milne (1882–1956) in 1926.

Zeno of Elea (490–460 BC) catalogued forty different paradoxes of motion, ultimately concluding that motion is logically impossible.

Comments on Selected Exercises

In the following comments, the formula reference denoted $(p.q)$, where p is in roman numerals and q is an arabic integer, corresponds to formula (q) in chapter p. In some exercises, we give computer algebra system expressions, written in *Mathematica*, which are adaptable to any other CAS. Any exercise marked with an asterisk is one that I cannot solve.

Chapter II

1. Let R be the radius and M be the mass of planet U. The volume of U is $2\pi R^3$. Let P be the point $(0, 0, s)$ in rectangular coordinates. In cylindrical coordinates the volume element is $r \, \Delta\theta \, \Delta r \, \Delta z$. Then by (II.1) the acceleration on a point-mass at P is

$$a_1(s) = \frac{GM}{2\pi R^3} \int_{-R}^{R} \int_{0}^{R} \int_{0}^{2\pi} \frac{(z-s)r}{(r^2 + (z-s)^2)^{\frac{3}{2}}} \, d\theta \, dr \, dz.$$

For $0 \le s \le R$, this integral simplifies to

$$a_1(s) = -\frac{GM}{R^3} \left(2s + \sqrt{R^2 + (R-s)^2} - \sqrt{R^2 + (R+s)^2} \right). \tag{1}$$

If $s > R$, the only change in (1) is that the $2s$ term is replaced by $2R$.

2. Let $P = (s, 0, 0)$ in rectangular coordinates. With respect to the **i** direction, the analog of (II.1) is

$$a(s) = \frac{GM}{2\pi R^3} \cdot \frac{x - s}{((x-s)^2 + y^2 + z^2)^{\frac{3}{2}}}.$$

Thus

$$a_2(s) = \frac{GM}{2\pi R^3} \int_{-R}^{R} \int_{0}^{R} \int_{0}^{2\pi} \frac{(r\cos\theta - s)r}{(r^2 - 2rs\cos\theta + s^2 + z^2)^{\frac{3}{2}}} \, d\theta \, dr \, dz. \tag{2}$$

305

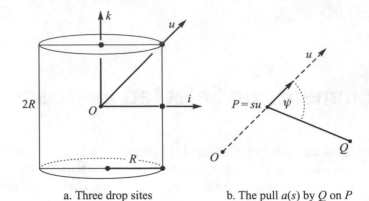

a. Three drop sites   b. The pull $a(s)$ by Q on P

Figure 1. Planet U, Exercises 1–3

To find a_3, let $P = s\mathbf{u}$ where $\mathbf{u}$ is the unit direction $\mathbf{u} = (1, 0, 1)/\sqrt{2}$. Figure 1b is the analog of Figure II.2. Thus

$$\cos\psi = \frac{(x - s/\sqrt{2},\, y,\, z - s/\sqrt{2}) \cdot (1, 0, 1)\sqrt{2}}{\sqrt{(x - s/\sqrt{2})^2 + y^2 + (z - s/\sqrt{2})^2}}.$$

With respect to the $\mathbf{u}$ direction, the analog of (II.1) is

$$a(s) = \frac{GM}{2\pi R^3} \cdot \frac{x/\sqrt{2} + z/\sqrt{2} - s}{((x - s/\sqrt{2})^2 + y^2 + (z - s\sqrt{2})^2)^{\frac{3}{2}}}.$$

Now use $x = r\cos\theta$ and $y = r\sin\theta$ to obtain $a_3(s)$, a triple integral similar to (2). Note: The gravitational acceleration at the rim of planet U may not be directed towards the origin. Nevertheless, down a hole drilled from the rim to the center, the effective gravitational acceleration will be $a_3(s)$.

3. Use (II.17) and (II.18). Remember that with respect to $a_3(s)$, the rock falls $R\sqrt{2}$ units.

4. The average density of the earth is $\delta = 5.52$ grams/cm^3. For the radius ρ of the earth, $2p = 6378$ km. Let R be the radius of the disk. Let $a_4(R)$ be the acceleration function for planet X at $(0, 0, p)$. Then by a modification of $a_1(s)$ from Exercise 1,

$$a_4(R) = \delta G \int_0^R \int_{-p}^p \int_0^{2\pi} \frac{(z - p)r}{(r^2 + (z - p)^2)^{\frac{3}{2}}}\, d\theta\, dz\, dr.$$

Then solve $a_4(R) = -9.8$ m/sec^2.

5. Assume that the density of air at $R = 5600$ km from the earth's center is that of the air at sea level, $\delta = 1.2$ kg/m^3. So $M = \delta V$ where V is the volume of a sphere of radius R. By (II.7), this mass of air induces an acceleration of $-GM/R^2$ on the inhabitants of Pellucidar.

6. As in Exercise 5, let $R = 5600$ km, and let the density of air on Pellucidar's surface be $\delta_R = 1.2$ kg/m^3. Assume that δ is 1.6 kg/m^3 at radial distance $R - 5$ km from Pellucidar's center. Now use (II.13) to determine the constants δ_0 and k.

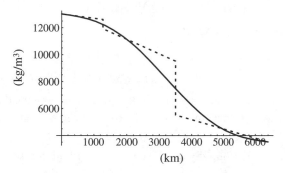

Figure 2. A curve approximating earth's density

7. From measurements of the figure, let

```
data = {{0, 13000}, {1275, 12600}, {1280, 11900}, {3500, 9500},
                {3505, 5500}, {6400, 3500.}}
```

The command

$$\text{fit[data, } \{1, \ x, \ \text{Sin}[\pi \ x/3200]\}, \ x]$$

returns

$$13024 - 1.48467x + 1179.4\sin(\pi x/3200),$$

whose graph is in Figure 2. Add more data points in data and more base elements to the function.

8. Modify (II.17) so that

$$v(s) = -\sqrt{2\int_R^s a\,dr + v_0^2},$$

where v_0 is the velocity at R.

9*.   Interpret the phrase *constant gravity* to mean that the magnitude of the sum of gravitational accelerations in the x, y and z directions is a constant. Orient the cube so that its center is at the origin and $\mathbf{k}$ is normal to its top face and $\mathbf{i}$ is normal to another face. To start this exercise, imagine that gravity is constant along the line segment $\mathcal{L} = \{(p, 0, 50)| -50 < p < 50\}$ on the top face. Let $\delta(x, y, z)$ be the density of the cube at (x, y, z). At $(p, 0, 50)$, the acceleration in the $\mathbf{k}$ direction is

$$a_{\mathbf{k}} = G \int_{-50}^{50} \int_{-50}^{50} \int_{-50}^{50} \frac{(z - 50)\delta(x, y, z)}{((x - p)^2 + y^2 + (z - 50)^2))^{\frac{3}{2}}} \, dx \, dy \, dz.$$

The acceleration in the $\mathbf{i}$ direction is

$$a_{\mathbf{i}} = G \int_{-50}^{50} \int_{-50}^{50} \int_{-50}^{50} \frac{(x - p)\delta(x, y, z)}{((x - p)^2 + y^2 + (z - 50)^2))^{\frac{3}{2}}} \, dx \, dy \, dz,$$

and the acceleration in the $\mathbf{j}$ direction is 0, since we may assume that $\delta(x, y, z)$ is symmetric with respect to the natural axes. Now find a function $\delta(x, y, z)$ so that $\sqrt{a_{\mathbf{i}}^2 + a_{\mathbf{k}}^2}$ is a constant for all p, $-50 \leq p \leq 50$. Try to extend this approach so that gravity is constant on the whole top face rather than just $\mathcal{L}$.

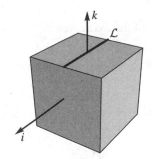

Figure 3. Planet Y, Exercise 9

10.   Use Case 3 of Chapter II. Recall that $k = GM$.

Chapter III

1.   Adapt (II.17) and Case II.1 with velocity v_0 at $r = R - 100$ miles, where R is the earth's radius:

$$v(s) = \sqrt{v_0^2 - \frac{g}{R}(s^2 - (R - 100)^2)}.$$

Then observe that $v(R) = 0$, and solve for v_0. Watch your units!

2.   Solve the system,

$$\alpha T = 14000 \quad \text{and} \quad \alpha T^2/2 = 700.$$

The weight of the astronaut is $180\alpha/g$ where $g = 32$ ft/sec^2.

3.   First find the balance point r_b between the earth and the moon. Then use (III.9) and solve for v_0:

$$83\frac{1}{3}\ \text{hr} = \int_R^{r_b} \frac{1}{v}\, ds.$$

4.   Let $s(t)$ be the distance traveled from time 0 to time t. Since the velocity is 0 at $t = 0$, then the mean velocity from 0 to t is $s'(t)/2$. With this notation, Merton's rule becomes

$$s(t) = \frac{s'(t)}{2} t \text{ which is } \frac{2}{t}\, dt = \frac{1}{s}\, ds,$$

which means that $\ln t^2 = \ln s + C$, which is $\ln(t^2/s) = C$, which is $s = t^2/e^C$, where C is a constant.

5.   Let $t = i\,\Delta t$. Then da Vinci's argument becomes

$$\frac{L(t)}{\Delta t} = \frac{s(t) - s(t - \Delta t)}{\Delta t}.$$

That is, $L(t)/\Delta t \approx s'(t)$. Therefore $E(t) = L(t + \Delta t) - L(t) \approx (s'(t + \Delta t) - s'(t))\Delta t$, which means that

$$\frac{E(t)}{(\Delta t)^2} \approx \frac{s'(t + \Delta) - s'(t)}{\Delta t} \approx s''(t) = g.$$

That is, $E(t) = E(i\,\Delta t) \approx (\Delta t)^2 g$.

6.   After firing the cannon, we can find the time lapse T to impact, and the range Q. From (III.4), recover v_0 and α. That is, solve the system $v_0 T \cos\alpha = Q$ and $gT = 2v_0 \sin\alpha$.

8.   The answer is about $56°$. This exercise came from Dou [25].

9. Use this Mathematica code to implement (III.12):

R=4000; μ=0.0123; g=32/5280; d=222000; L=1082; v₀=6.89;

$$v[r_-] := \sqrt{2g\,R^2(\tfrac{1}{r} + \tfrac{\mu}{d-r}) + v_0^2 - 2g\,R^2(\tfrac{1}{R} + \tfrac{\mu}{d-R})}$$

Then v[d-L]= 1.43 mi/sec. We assume that Columbia just misses the moon. The escape velocity of the moon is 1.48 mi/sec. So the moon is just barely able to keep Columbia from escaping.

10. Use (III.10) where $\mu = 1/2$ and D is one billion miles. M is the mass of the sun. A good solution will be a function giving the travel time in terms of v_0.

Chapter IV

1. Use the same trick as in (VIII.26) to write t as a function for all θ. That is,

$$t(\theta) =$$

$$\frac{h^3}{k^2(1-e^2)^{\frac{3}{2}}}\left(\theta - 2\tan^{-1}\left(\frac{e\sin\theta}{1+\sqrt{1-e^2}+e\cos\theta}\right) - \frac{e\sqrt{1-e^2}\,\sin\theta}{1+e\cos\theta}\right),$$

$$(3)$$

With $e = 0.017$, earth's eccentricity, define

$$T(\theta) = \frac{365\,t(\theta)}{t(2\pi)}$$

to sidestep dealing with values of h and k. To find θ on April 1, solve $90 = T(\theta)$, since April 1 is 90 days after January 1.

2. With $e = 0.5$, use $T(\theta)$ as defined in Exercise 1. Spring occurs on day $T(\pi/2)$.

3*. Try working in three dimensions. The initial rotational velocity is $2\pi R \cos(28°)/T$.

4. Try this Mathematica code.

μ=0.0123; g = 32/5280(60 60 24)²; R=4000; ρ=222000; t₀ = 2.6;
Q=27.3;
$\omega[t_-] := 2\pi/Q\,(t - t_0)$;
$L[t_-] := \sqrt{(\rho\cos[\omega[t]] - r[t]\cos[\theta[t]])^2 + (\rho\sin[\omega[t]] - r[t]\sin[\theta[t]])^2}$;
verne=NDSolve[{r[t]θ''[t] + 2r'[t]θ'[t]
        == $\mu\,g\,\rho\,R^2/L[t]^3(-\cos[\omega[t]]\sin[\theta[t]] + \sin[\omega[t]]\cos[\theta[t]])$,

```
r''[t] − r[t](θ'[t])² == −g R²/r[t]² + μ g R²/L[t]³(ρ Cos[ω[t]]Cos[θ[t]]
    +ρ Sin[ω[t]]Sin[θ[t]] − r[t]),
r[0] == R, θ[0] == 0, r'[0] == 6.9 60 60 24,
    θ'[0] == 2π}, {r[t], θ[t]}, {t, 0, 3}];
rV=(r[t]/.verne)[[1]];  θV=(θ[t]/.verne)[[1]];
moon[t_]:=Graphics[Disk[ρ{Cos[ω[t]], Sin[ω[t]]}, 1000]];
track=ParametricPlot[rV{Cos[θV], Sin[θV]}, {t, 0, 2.85}];
Show[track, moon[2.85], PlotRange→{{0, ρ+1000}, {0, 14500}}];
```

Implementing this code produces Figure IV.5. The time 97 hours and 13 minutes is about 4.05 days. Experiment with the code above. Change t_0 and the initial condition 6.9 mi/sec^2 so as to achieve moon-rendezvous in 4.05 days rather than the 2.85 days of the text. You'll need to change the interval over which NDSolve is generating the solution, from {t, 0, 3} to {t, 0, 4.1}, and change the PlotRange values appropriately.

5. The moon's eccentricity is $e = 0.055$ and the (sidereal) period is 27.3 days. Therefore use the function,

$$T(\theta) = \frac{27.3\, t(\theta)}{t(2\pi)},$$

where $t(\theta)$ is (3) of Exercise 1 above.

6. Using the function T[θ] from Exercise 5 generate a data set.
    data = Table[{T[θ], θ}, {θ, 0, 2π, 0.1}];
    Now fit a function $\theta = p(t)$ as a linear combination of 1, t, and $\sin(2\pi t/27.3)$ through data.
    p=Fit[data, {1, t, Sin[2π t/27.3]}];
    Implementing the above code generates the function

$$p(t) = 0.00437665 + 0.229828t + 0.107262 \sin(0.230153t).$$

The graphs of $p[t]$ and the parametric plot of {T[θ], θ} are almost identical; both follow the curve as given in Figure 4, giving t in days versus θ in radians.

7. Use the code of Exercise 4. Change the function $\omega[t]$ to

$$\omega[t_-] := p[t − t_0]$$

where $p(t)$ is the function derived in Exercise 6.

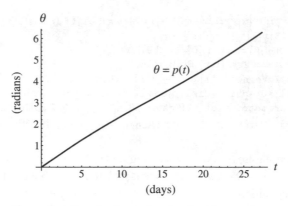

Figure 4. A function approximating θ with respect to t

8. The barycenter, or the center of gravity, of the earth-moon system is 1067 miles beneath the earth's surface. Assume that the earth and the moon rotate in circles about this point with period 27.3 days. Let $E(t) = 2933(\cos(2\pi/27.3), \sin(2\pi/27.3))$ be the motion of the earth. Obtain a function $N(t)$, analogous to $L(t)$, giving the distance of Columbia from the earth's center. Then obtain two non-zero functions $A_{E,\theta}$ and $A_{E,r}$ analogous to $A_{M,\theta}$ and $A_{M,r}$, ultimately giving a system analogous to (IV.20) except that the right-hand side of the first equation is $A_{E,\theta} + A_{M,\theta}$.

9. Let

$$E(t) = 93000000(\cos(2\pi t/365), \sin(2\pi t/365))$$

and let

$$M(t) = E(t) + 222000(\cos(2\pi/27.3), \sin(2\pi/27.3)),$$

be the motion of the earth and the moon about the sun. Adapt (IV.20) just as in Exercise 8. This time the right-hand side of the second equation will be $A_{S,r} + A_{E,r} + A_{M,r}$.

10. This exercise is a repeat of Exercise 8. Find the barycenter by remembering that $M_1 d_1 = M_2 d_2$ where M_1 and M_2 are the two masses and d_1 and d_2 are the distances of the respective masses from the barycenter.

Chapter V

1. Use (V.7) and the inverse square law for gravitational acceleration above the earth's surface.

2.   Use (V.7) and a linear relation for gravitational acceleration within the earth.

3.   Since the mass of the system is entirely at the end of the ladder, then the moment of inertia around the pivot is $I = mr^2$, and the downward force at the end of the ladder is $mg \cos \theta$. That is $mr^2(-\theta'') = mgr \cos \theta$, which simplifies to

$$\theta'' = -\frac{g}{r} \cos \theta.$$

4.   By similar triangles the height z of the midpoint of the ladder is always $\frac{1}{2}y$, where y is the height of the ladder's tip. Thus $z'' = \frac{1}{2}y''$. So when the tip of the ladder is experiencing an acceleration of $3g/2$ the midpoint is experiencing an acceleration of $3g/4$.

5.   Concentrate all the mass at the midpoint of the ladder. Then $y_c'' = -2g$.

6.   Use the code,
```
α = 1/(88π); q = 176π;
poe = NDSolve[{α t θ''[t] + 2α θ'[t] + 32Sin[θ[t]] == 0,
   θ[176π] == 0, θ'[176π] == 1},
   θ[t], {t, q, 16q}, MaxSteps → ∞][[1, 1, 2]];
ω[s_]:= poe/.t→s;
x[t_]:=α t ω[t]; v[t_]:=α ω[t]+α t ω'[t];
Plot[v[t], {t, q, 16q}];
```
In the above code, note that $v(t)$ is the derivative of $x(t)$, where $x(t)$ is the arclength from the pendulum to the vertical through the pivot. Implementing the above code will generate the graph of $v(t)$. Be prepared for a long runtime. A quicker way is to use Bessel functions: use (V.22), and then plot α $\theta[t]+\alpha$ t $\theta'[t]$.

7.   Suppose that the ladder shrinks at c feet/second. We shall assume that the density of the ladder is always uniform. Then the length of the ladder is $r = 40 - ct$. The analog to (V.8) is

$$y = \frac{cr - kx}{y}.$$

Equation (V.9) remains valid, except that r varies with time t.

8*.   One approach is to find all those times t at which the (unpumped) pendulum reaches an extreme amplitude on the right. Then, since the period of a (non-descending) pendulum is independent of amplitude, pump the pendulum at each of these times using a constant multiple of the Dirac delta function, as found in the back of a typical text on differential equations.

9. For ideas, see Wirkus [**98**].

10. For ideas, see Oprea [**60**].

Chapter VI

1. Solve $s(12) = 75$ for κ where $s(t)$ is (VI.12). The Mathematica syntax for such number crunching (provided $s(t)$ is defined in Mathematica) is

 FindRoot[s[12]==75, {κ, p}]

where p is an initial guess as to the solution.

2. This exercise is a repeat of Exercise 1. Solve $s(50) = 2000$ for κ.

4. The differential equation for the ball's descent is

$$Mv' = -mg + 4\pi k v^2,$$

where M is the mass of the ball, m is the apparent mass of the ball, $g = 32$ ft/sec^2, and k is a constant to be determined. Rewrite this differential equation as

$$v' = \frac{4\pi k}{M}\left(v^2 - \frac{mg}{4\pi k}\right).$$

Terminal velocity v_T occurs when

$$v_T = \sqrt{\frac{mg}{4\pi k}}.$$

Solve the equation

$$v_T^2 = \frac{mg}{4\pi k} = 100^2$$

for k, where $m = (15.1 - 2)\frac{4}{3}\pi 2^3$. Then use this value of k to determine κ for Maria and her sinker.

5. The apparent mass of Maria is $m_\mu = -263$ slugs while the apparent mass of the sinker is $m_\sigma = 7634$ slugs. When Maria is at the surface, held up by a crane, with the sinker deployed, a force of $m_\sigma g \approx 244288$ pounds goes through the tether. If this tether has cross section 1 in^2, then the tension in the tether is 244288 pound/in^2, strong enough to snap any normal rope. As in (VI.3), when the tandem spheres descend and reach terminal velocity, Maria tries to go up with a force of $m_\mu g$ and experiences a drag of $\pi k r^2 v^2$, where $\pi k r^2 = \kappa \approx 2.31$ slugs/ft and $v \approx -226$ ft/sec, all of which gives a tension of about $126, 400$ pounds/in^2.

6. From the example in the text, $\kappa = 2.31$ slugs/ft. Recall that $\kappa = \pi k R^2$ where $R = 4.5$ feet. So $k = 0.0363$. The new value of κ is $\kappa_{new} = \pi k r^2$ where r is the outer radius of the fin. Use (VI.9) and solve the equation $-m_\mu g/\kappa_{new} = 100$ for r, (where $m_\mu = -263$ slugs). (The water turbulence associated with this design probably makes this idea impractical.)

8. Maria's apparent mass is $m_\mu = -263$ slugs. At time $t = c$, allow Maria to take on water at ρ ft^3/sec. Then Maria's new apparent mass for $t \geq c$ is the function $m_{new}(t) = m_\mu + 2\rho(t - c)$ slugs. Solve (VI.8), starting at $t = c$, (knowing $v(c)$ and $s(c)$) by replacing m_μ with $m_{new}(t)$. Be careful about reversing Maria's buoyancy! Don't let her sink.

Chapter VII

1. Let $A = R(\cos 28° \cos 24°, \cos 28° \sin 24°, -\sin 28°) \approx (5162.3, 2298.4, -3004.6)$ and $B = R(\cos 52°, 0, \cos 52°) \approx (3940.2, 0, 5043.3)$ be the coordinates of Kimberly and London respectively. Thus $\gamma \approx 4229.2$ km, and $2\mu/\lambda \approx 6.7$ km; that is, the letter turns back 6.7 km short of London. Now interchange A and B so that A has the coordinates of London and B has the coordinates of Kimberly. To find the time at which the letter is at Kimberly, solve $s(t) = -\gamma$, where $s(t)$ is (vii.7). The Mathematica code is

 `FindRoot[s[t]==−`γ`, {t, 40 60}]`

We supplied `FindRoot` with an initial guess of forty minutes (since the root should be about half the period). To this command, Mathematica returns $t \approx 2493.7$ sec. To find the velocity of the letter at this time, use

 `s'[2493.72]`

and get -0.29. That is, the speed at which the letter lands in Kimberly from London is 0.29 km/sec.

2. An easy way to solve this problem is to start the train with zero velocity at the south pole and determine its velocity v_e at Entebbe. Then if the train is started at Entebbe with velocity $-v_e$, the train should arrive at the south pole with zero velocity.

3. If A and B have the same third component, then $u_1^2 + u_2^2 = 1$, which means that the corresponding period of $2\pi/\sqrt{\lambda}$ is the maximum value, namely $2\pi/\sqrt{k - 4\pi^2/Q^2}$. Contrariwise, if A and B fail to lie on the same latitude, then $u_1^2 + u_2^2 < 1$, giving rise to a period less than $2\pi/\sqrt{k - 4\pi^2/Q^2}$.

4*. A good start on this problem is to begin reading geophysics texts, such as Lowrie [**51**].

5. This is a start in tackling Exercise 10.

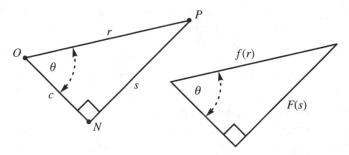

Figure 5. Gravitational attraction along the chord, (Exercise 6)

6. Let $f(r)$ be the gravitational strength of the earth at r units from O. The main problem in this exercise is finding the gravitational strength $F(s)$ along the chord. To do so, refer to Figure 5. Observe that $\theta = \tan^{-1}(s/c)$. Furthermore, since $\sin\theta = s/r$ then $r = s/\sin(\tan^{-1}(s/c))$. By similar triangles, $f(r)/F(s) = r/s$. Therefore $F(s) = \sin(\tan^{-1}(s/c)) f(s/\sin(\tan^{-1}(s/c)))$.

8. This is the celebrated brachistochrone problem. See Haws [**38**] and Johnson [**43**], for example.

9. See Lin [**48**].

10. See Figure VIII.7 for a curved arc through the equatorial plane which has period $\sqrt{k}$, for example. Are there faster paths between the endpoints of this curve? No, in fact these arcs are hypocycloids!

Chapter VIII

1*. As a pebble falls towards the center of the planet, its speed increases. Thus a pebble falling from its apogee reaches its perigee in a finite time. By radial symmetry of f, the pebble rises by reversing the way that it fell. Can you make this argument rigorous?

2.* Here's a way to generate the speed of a field:
```
pebble=NDSolve[{z"[θ]+z[θ]==2/z[θ]^4, z[0]==1,
z'[0]==0}, z[θ], {θ, 0, 2π}][[1,1,2]];
f[s_]:=1/pebble/.θ→s
FindRoot[f'[θ],{θ, π}]
```
In response to these commands, Mathematica returns $\theta \to 2.8917$. So the speed of f_1 as given in the exercise is about 0.89737.

10*. My physicist friends assure me that the similarity between the spirographic structure of the nebula and the gravitational pebble-paths or pebbleholes is coincidental.

Chapter IX

1. By (IX.9), Γ is an arc of the circle

$$\left(x - \frac{\sqrt{1-c^2}}{c}\right)^2 + (y+1)^2 = \frac{1}{c^2},$$

since $a = 1 = b$. Since the midpoint of the segment from the epicenter at $(0, 0)$ to Helen at $(2, 0)$ is $(1, 0)$, then $\sqrt{1-c^2}/c = 1$, which means that $c = 1/\sqrt{2}$. Thus the deepest descent y_{max} of the signal is $100(\sqrt{2} - 1)$ miles. To find the time lapse along a signal, recall that time is distance divided by rate. The distance along the arc of a signal is given by $\sqrt{1 + (\frac{dx}{dy})^2}$ and the rate is v, thus the time lapse T along a signal is

$$T = 2\int_0^{y_{max}} \frac{\sqrt{1 + (\frac{dx}{dy})^2}}{v}\,dy,$$

which via (IX.6) simplifies to

$$T = 2\int_0^{y_{max}} \frac{1}{v\sqrt{1-(cv)^2}}\,dy. \tag{4}$$

Thus by (4), the time lapse for the signal which Helen heard is

$$T = 2\int_0^{\sqrt{2}-1} \frac{1}{(1+y)\sqrt{1-(1+y)^2/2}}\,dy \approx 1.762 \text{ minutes},$$

so the blast occured at 9:58:15 am.

2. Let θ be a function in terms of r. We need to find the arc-length along the polar curve $(r, \theta(r))$. To do so, let $x = r\cos\theta$ and $y = r\sin\theta$. Then $x' = \cos\theta - r\theta'\sin\theta$ and $y' = \sin\theta + r\theta'\cos\theta$, where all derivatives are with respect to r. Thus the arc-length along the curve from r_1 to r_2 is

$$\int_{r_1}^{r_2} \sqrt{(x')^2 + (y')^2}\, dr = \int_{r_1}^{r_2} \sqrt{1 + r^2(\theta')^2}\, dr. \tag{5}$$

By (5), the time T for a signal to go along this polar curve is

$$T = \int_{r_1}^{r_2} \frac{\sqrt{1 + r^2(\theta')^2}}{v}\, dr.$$

Now replace θ' in the above integral with (IX.19), and simplify.

3. 946 seconds or 15.8 minutes.

5. Consider the Mathematica code:

```
R=6400; c=320; a[b_]:=10.- b R; r0[b_]:= a[b]
  c/(1- b c); v[r_, b_]:=a[b]+ b r;
```

$$\Theta[\text{b\_}] := \text{Re}[2 \int_{r0[b]}^{R} \frac{c\, v[r, b]}{r\sqrt{r^2 - c^2 v[r, b]^2}}\, dr] \cdot \frac{180}{\pi}$$

Figure IX.14 shows the graph of $\Theta(b)$. By trial and error, $\Theta(-0.000837) \approx 90.01°$. Thus $r0(-0.000837) \approx 3876$. That is, the core-mantle boundary is about 3876 km from the earth's center.

6. The central angle at which f is a minimum is 0.3632 radians, so that $f(0.3632) \approx 2.87$. $P \approx (-1.67, 0.56)$. The time from $A = (2, 0)$ to P along Γ_1 is 1.08 units, and along Γ_2 is 1.24 units.

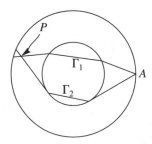

Figure 6. Two paths to P, (Exercise 6)

7. When generating paths which lie entirely in the mantle, such as Γ in Figure 7a, (IX.19) (with the negative option) will give $\theta(r)$ for $r_{min} < r \leq 2$, where $r_{min} = 5c/(1 + c)$. Since the differential equation is undefined at r_{min}, generate $\theta(r)$ from $r_{min} + \delta$ to 2, for some sufficiently small δ value such as 0.0000001. To generate the second half, define $\omega(r) = 2\theta(r_{min} + \delta) - \theta(r)$. To obtain the entire path, plot both $r(\cos\theta(r), \sin\theta(r))$ and $r(\cos\omega(r), \sin\omega(r))$.

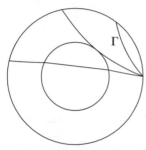 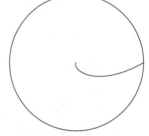

a. Three paths, (Exercise 7)   b. Spiraling to the center, (Exercise 8)

Figure 7. Signal paths

8. In this case, (IX.16) means that $c = 2$. Using the positive option in (IX.19) gives the differential equation $\theta' = 1/(r\sqrt{3})$. So $\theta = \ln(r/2)/\sqrt{3}$, whose polar graph is shown in Figure 7b. How long does it take for the signal to travel this curve?

10. π

Chapter X

2. To prove this observation, use the trigonometric identities $2\cos(\alpha)\cos(\beta) = \cos(\beta+\alpha)+\cos(\beta-\alpha)$ and $2\cos(\alpha)\sin(\beta) = \sin(\beta+\alpha) + \sin(\beta - \alpha)$. Thus the polar curve $r = \cos(m/n\,\theta)$ is the rectangular parametrized curve $(\cos(m/n\,\theta)\cos\theta, \cos(m/n\,\theta)\sin\theta)$. With $\alpha = m/n\,\theta$ and $\beta = \theta$, the parametrized curve is $\frac{1}{2}(\cos(\theta + m/n\,\theta) + \cos(\theta - m/n\,\theta),$ $\sin(\theta+m/n\,\theta)+\sin(\theta-m/n\,\theta))$. Then, let $t = n\,\theta$, giving the parametrized curve $\frac{1}{2}(\cos((n+m)t) + \cos((n-m)t), \sin((n+m)t) + \sin((n-m)t))$. To graph these curves for Exercises 1–5 use the Mathematica code:

```
T[p_, q_, λ_] := ParametricPlot[{λ Cos[p t] + (1 − λ)Cos[q t],
    λ Sin[p t] + (1 − λ)Sin[q t]}, {t, 0, 2π},
      AspectRatio → Automatic, Axes → None];
```

For example, the first curve in Figure 8 is generated by the command T[1 + 2, 1 − 2, 1/2].

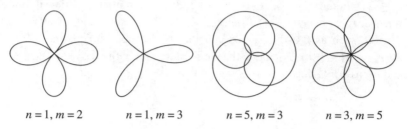

$n = 1, m = 2$ $n = 1, m = 3$ $n = 5, m = 3$ $n = 3, m = 5$

Figure 8. Gallery of polar flowers, (Exercise 2)

7. Here's some code to generate parametric representations of $F^{(n)}(t) = 0 = F^{(m)}(t)$.
F[t_] := x(Sin[p t] − Sin[q t]) + y(Cos[q t] − Cos[p t]) + Sin[q t − p t];

$$\text{Solve}[\{F[t] == 0, F'[t] == 0\}, \{x, y\}] \tag{6}$$

The output from (6) is

$$\{\{x \to (q\,\text{Cos}[p\,t] + p\,\text{Cos}[q\,t])/(p + q),$$
$$y \to (q\,\text{Sin}[p\,t] + p\,\text{Sin}[q\,t])/(p + q)\}\}$$

To get the parametric representations of $F^{(n)}(t) = 0 = F^{(m)}(t)$, adapt (6), using n prime marks (′) and m prime marks (′) instead of 0 and 1 mark, respectively.

Figure 9. A fractional envelope, $u = 1.3$, $p = 4$, $q = 1$, (Exercise 8)

8. Here's some code to help simplify the formulas.
F[u_] := x(p^u Sin[p t + π u/2] − q^u Sin[q t + π u/2]) + y(q^u Cos[q t + π u/2] − p^u Cos[p t + π u/2]) − (p − q)^u Sin[p t − q t + π u/2];

```
x1 = Solve[{F[1.3] == 0, F[2.3] == 0}, {x, y}][[1, 1, 2]]
y1 = Solve[{F[1.3] == 0, F[2.3] == 0}, {x, y}][[1, 2, 2]]
```
Thus $x1$ and $y1$ will be parametric representations of the desired envelope for $u = 1.3$. (These formulas are fairly long.) To plot this envelope (whose graph is Figure 9) use
```
p = 4; q = 1; ParametricPlot[{x1, y1}, {t, 0, 2π},
    AspectRatio → Automatic, Axes → None];
```

Chapter XI

1. If Cavor is launched at $t = 0$, then it intersects the moon's orbit at the point $P = R(1, \sqrt{3599})$, where $R = 6400$ km. The angle between $\mathbf{i}$ and P is $\theta_m = \tan^{-1}\sqrt{3599}$. It takes the moon $t_m = 6.82$ days to traverse central angle θ_m from the earth. The difference between 9.5 days and t_m days is $t_n = 2.67$ days. So if the moon was 2.67 days from reaching angle $\theta = 0$ when Cavor is launched (at angle $\theta = 0$), moon rendezvous occurs in 9.5 days. The central angle between the moon at this launch time and the moon at rendezvous time is therefore $\theta_n = \theta_m + 2\pi\, 2.67/27.3 \approx 2.185$. Now let $t = 0$ be the time at which earth's center, Cavor and the moon are on the same line. Let x be the launch time. Solve the equation $2\pi x - 2\pi x/27.3 = \theta_n$, getting $x \approx 0.3610$ days. Thus if $t = 0$ is midnight, then one solution for a moon rendezvous is to launch Cavor at 8:40 am.

3. Observe that the point $(5.5, 0.3)$ appears to be on the curve. This point corresponds to a rendezvous time of 5.2 days.

5. Here's an intuitive argument which can be made rigorous. The motion of the moon around the earth has a greater effect in returning to earth than does the motion of the moon about its axis. Therefore, the best strategy is to remove as much tangential orbital velocity from Cavor's launch site as possible. This occurs at $\omega = \pi$.

6. Here's the required Mathematica code.
```
g = 9.8/1000(24 60 60)^2; Q = 27.3; rd = 384000; re = 6400;
k = g re^2; rm = 1738;
L = rd − rm; vx = 0; vy[q_] := 2π rm(−1)/q + 2π L/Q;
h[q_] := L vy[q];
a[q_] := −Abs[1/L − k/(h[q]^2)]; φ = 0;
```

$$r[q_-, w_-] := 1/(a[q]Cos[w] + k/h[q]^2); \tag{7}$$

Implementing r[.01Q, π] gives 65,785.5; that is, Cavor's nearest approach to the earth's center is 65,785.5 km.

To find q at which the apogee for (7) is r_e, use FindRoot[r[q, π] == r_e, {q, 0.01Q}] and obtain {$q \to 0.151856$}, which means that the moon would need to rotate about its axis once every 0.15 days in order for Cavor to make it home.

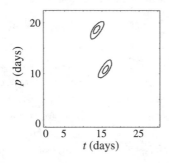

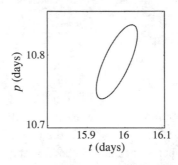

a. Contours at 200,000 km and 100,000 km   b. Contour at 6400 km

Figure 10. A Wellsian solution (Exercise 8)

8. Here is some useful Mathematica code:

```
rs = 150000000; q = 27.3; rd = 384000;
e[t_] := rs {Cos[2π t/365], Sin[2π t/365]};
S[t_] := e[t] + rd {Cos[2π t/q], Sin[2π t/q]};
c[t_, p_] := S[p] + S'[p](t − p);
d[t_, p_] := √((e[t] − c[t, p]).(e[t] − c[t, p]))
```

To search for the time at which Cavor should cloak itself to gravity, contour-plot $d(t, p)$. In particular, we consider a coarse time frame where $0 < t < 30$ and $0 < p < 22$ days. Now plot the level curves, $d(t, p) = 100000$ and $d(t, p) = 200000$, as shown in Figure 10a. Note that there is a spurious solution at about (15, 20). The code to generate these contours is

```
ContourPlot[d[t, p], {t, 0, 30}, {p, 0, 22},
    Contours → {100000, 200000},
    ContourShading → False, PlotPoints → 50];
```

Now zoom in on Figure 10a at about (15, 10) to obtain 10b, whose lone contour corresponds to distance 6400 km. As can be seen, one solution is (16, 10.8). That is, Cavor should cloak itself on day 10.8 and arrive at earth on day 16. That is, it takes about 5.2 days to journey home.

Chapter XII

1. Let v_e be the escape velocity. By (XII.8), $v_e = \sqrt{2gR}$. Solving $155 = \sqrt{2 * 32(60 * 60)^2/5280R}$ for R gives $R \approx 551$ miles.

2. Suppose that the ball is thrown from point $P = (a, 0, b)$ on the asteroid. On the nonrotating asteroid, let $\mathbf{v}_p$ be the initial velocity of the ball corresponding to a particular trajectory Γ from P to the north pole. Γ is an ellipse lying in a plane through P, the north pole, and the origin. Now if a ball is thrown from P to the north pole on a rotating asteroid, its trajectory must be an ellipse lying in the plane of P's initial location, the north pole and the origin. Thus, if we wish the trajectory of the ball on the rotating asteroid to be Γ, then the initial velocity of the ball is $\mathbf{v}_p - \omega a\mathbf{j}$, where $\omega = 2\pi/\text{min}$, (assuming the asteroid rotates counterclockwise). That is, for the ball to reach the north pole, its initial velocity must effectively nullify the tangential velocity which the ball inherits from the rotating asteroid. Therefore, the flight times are the same, but the initial velocity to achieve this end is greater for the rotating asteroid.

3. Throwing from B to A is the same as throwing from A to the south pole. Let $C = (0, -R)$ be the south pole. The analog for (XII.14), the initial velocity in terms of flight time τ for the ball to go from A to C, is

$$p = \frac{R}{\tau}(\sin(\omega\tau) - 1) \quad \text{and} \quad q = -\frac{R}{\tau}(\cos(\omega\tau) + \omega\tau).$$

Figure 11 gives the trajectory corresponding to a flight time of $\tau = 5$. Contrast this figure with Figure XII.8. With $\tau = 5$ sec, a throw from A to B requires velocity $(-3.5, -14.5)$, a speed of 14.9 ft/sec, whereas a throw

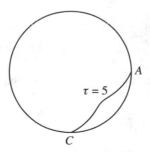

$\tau = 5$

A

C

Figure 11. Throwing to the south pole, (Exercise 3)

from A to C requires velocity $(-3.5, -7.5)$, a speed of 8.3 ft/sec. In general, it is easier to throw from B to A than from A to B.

4*.   Suppose that R, ω, f[t, p, q], M[t] and F[t, p, q] are all defined in Mathematica according to (XII.11) and (XII.15). Try this code:

```
a = Show[Graphics[Circle[{0, 0}, R]], AspectRatio →
   Automatic];
P[τ_] := −R/τ(1 + Sin[ω τ]); Q[τ_] := R/τ(Cos[ω τ] − ω tau]);
pic[τ_] := Block[{z}, z = ParametricPlot[F[t, P[τ], Q[τ]], {t, 0, τ},
DisplayFunction → Identity];
Show[a, z, DisplayFunction → $DisplayFunction]];
Do[pic[τ, {τ, 11, 12, .25}]
```

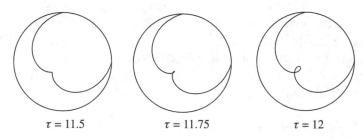

$\tau = 11.5$ $\tau = 11.75$ $\tau = 12$

Figure 12. Birth of a loop, (Exercise 4)

As Figure 12 shows, somewhere between time $\tau = 11.5$ and $\tau = 12$ sec, a loop develops.

5.   Here's the code that generates Figure XII.11b, (in color!). Modify it appropriately.

```
ω = 2π/10; M[t_] := {{Cos[ω t], Sin[ω t]}, {−Sin[ω t], Cos[ωt]}};
flag[x_, y_] := If[x > 0, If[y > 0, RGBColor[0, 1, 0],
   RGBColor[0, 0, 1]],
If[y > 0, RGBColor[1, 0, 0], RGBColor[0, 0, 0]]];
τ[x_, y_] := −2x/(x² + (y + w)²); good[z_] := flag[Part[z, 1],
   Part[z, 2]];
myC[x_, y_] := {good[M[τ[x, y]].{1. + τ[x, y]x, τ[x, y](y + w)}],
   Point[{x, y}]};
oh = Block[{bag, i, j, n, x, y}, n = 200;
   bag = {Pointsize[.015]};
Do[Do[bag = Append[bag, myC[−i/n, −j/n]], {i, 1, n}], {j, 1, n}];
bag; Show[Graphics[bag], AspectRatio → Automatic]];
```

The last command, oh, generates Figure 11b. When running this code, you may wish to use a smaller value of n; try n = 20 for a start.

6. Observe that

$$\lambda S_1(t) + \mu S_2(t) = t M(t) \left(\lambda \begin{bmatrix} p_1 \\ q_1 \end{bmatrix} + \mu \begin{bmatrix} p_2 \\ q_2 \end{bmatrix} \right).$$

So the line of symmetry is in the

$$\lambda \begin{bmatrix} -q_1 \\ p_1 \end{bmatrix} + \mu \begin{bmatrix} -q_2 \\ p_2 \end{bmatrix}$$

direction.

7. Let

$$G(t) = M(t) \begin{bmatrix} 1 + tp \\ tq \end{bmatrix}.$$

By (XII.18) with $R = 1$ and $\omega = 0$, when a ball is thrown with initial velocity (p, q), the flight time τ for it to go from $(1, 0)$ to wherever it strikes the wall of the unit circle is $\tau = -2p/(p^2 + q^2)$. Hence the ball strikes the unit circle at $G(\tau)$. Thus the line of symmetry for the $G(t)$ is the perpendicular bisector of the line with endpoints $G(0) = (1, 0)$ and $G(\tau)$.

8. Use (XII.18) with $\omega = 0$.

Chapter XIII

2. If P is the vertex of an ellipse at the semi-minor axis, $\overline{PF_1}$ has the same length as $\overline{PF_2}$.

3. The orothotomic of an ellipse with respect to one of its foci is a circle whose center is the other focus.

4. Here's the code used to generate the envelope for the polar curve $f(\theta) = 1/(2 + \cos\theta)$ with the beacon (of rotation rate ω) at the origin. In the code, we use L, n, c, s, W for the functions $\|g'\|$, $\mathbf{N}$, $\cos\alpha$, $\sin\alpha$, $\mathbf{W}_1$.

```
f[θ_] := 1/(2 + Cos[θ]);
g[θ_] := f[θ]{Cos[θ], Sin[θ]};
x[θ_] := f[θ]Cos[θ];
```

```
y[θ_] := f[θ]Sin[θ];
L[θ_] := √(x'[θ]² + y'[θ]²);
T[θ_] := {x'[θ], y'[θ]}/L[θ];
n[θ_] := {-y'[θ], x'[θ]}/L[θ];
κ[θ_] := (x'[θ]y''[θ] - x''[θ]y'[θ])/L[θ]³;
c[θ_] := If[(1/ω + f'[θ])/L[θ] < 1, (1/ω + f'[θ])/L[θ], 1];
s[θ_] := √(1 - c[θ]²);
W[θ_] := c[θ]T[θ] + s[θ]n[θ];
α[θ_] := ArcCos[c[θ]];
Q[θ_] := L[θ]s[θ]/(κ[θ]L[θ] + α'[θ]);
P[θ_] := g[θ] + Q[θ]W[θ]
```

Then use the commands

```
ω = 8; ParametricPlot[{{g[θ], P[θ]}, {θ, 0, 2π},
    AspectRatio → Automatic, Axes → None];
```

to generate figure XIII.13 with $\omega = 8$. To adapt this code for the polar curve $r = 1/(3 + \cos\theta)$, simply change the first command of the above code to `f[θ_] := 1/(3 + Cos[θ]);` For example, with $\omega = 10$, this code generates Figure 13.

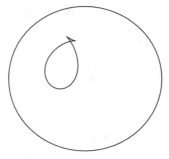

Figure 13. A fatter fish, $r = 1/(3 + \cos\theta)$, $\omega = 10$ (Exercise 4)

5. The Bivens transform of the ellipse $r = 1/(2 + \cos\theta)$ is the ellipse $r = 3/(5 + \cos\theta)$. Magnifying the ellipse by a factor of 2 gives the ellipse $r = 6/(5 + 4\cos\theta)$. At $\theta = 0$ and $\theta = \pi$, $r = 2/3$ and $r = 6$. Since the bisector curve of this ellipse is the fish curve, then two of the points on the y-axis are $(0, -1/3)$ and $(0, 3)$.

To find the coordinates of the cusps on the fish's tail, consider the parametrization of the envelope (XIII.28). To the code in Exercise 4 add these functions where $\alpha 1[\theta, \rho]$ and $Q1[\theta, \rho]$ represent the first partials of α and Q with respect to θ.

c[θ_-, ρ_-] := If[$\rho + f'[\theta]/L[\theta] < 1, \rho + f'[\theta]/L[\theta]$, 1];
s[θ_-, ρ_-] := $\sqrt{1 - c[\theta, \rho]^2}$;
W[θ_-, ρ_-] := c[θ, ρ]T[θ] + s[θ, ρ]n[θ];
α[θ_-, ρ_-] := ArcCos[c[θ, ρ]];
α1[θ_-, ρ_-] := D[α[w, ρ], w]/.w $\to \theta$];
Q[θ_-, ρ_-] := L[θ]s[θ, ρ]/($\kappa[\theta]$L$[\theta]$ + α1[θ, ρ]);
Q1[θ_-, ρ_-] := D[Q[w, ρ], w]/.w $\to \theta$;
P[θ_-, ρ_-] := g[θ] + Q[θ, ρ]W[θ], ρ;
P1[θ_-, ρ_-] := D[P[θ, p], p]/.p $\to \rho$;
Plotting P1[θ, 0] from $0 \le \theta \le 2\pi$ generates Figure XIII.20.
A cusp in P1 occurs when the derivative of P1[θ, ρ] with respect to θ is
0 at $(\theta, 0)$. By (XIII.30), this condition is satisfied when the coefficient of
W_1 is 0. Define this coefficient as $\beta(\theta)$, namely,

β[θ_-] := (D[Q1[θ, ρ], ρ] $-$ Q[θ, 0]D[$\alpha[\theta, \rho$], ρ]($\kappa[\theta]$L$[\theta]$
$\quad + \alpha$1[θ, 0])/.$\rho \to 0$.

Then use Newton's method on β to find a root.

FindRoot[$\beta[\theta]$, {θ, 1.}]

The above command returns {$\theta \to 0.895665$}, which means that the cusp is
at P1[0.895665, 0] $\approx$ (1.69135, 3.1875).

6. With the code of Exercise 5, the second Taylor term is given by
P2[θ_-, ρ_-] := D[P[θ, p], {p, 2}]/.p $\to \rho$;
Write the analagous function P3 to generate the third Taylor term whose
graph is given in Figure 14.

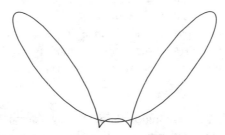

Figure 14. The third Taylor term, $r = 1/(2 + \cos \theta)$, (Exercise 6)

8. Use the same code as in Exercise 4, except replace the first command
with f[θ_-] := 1 + Cos[θ];
Since the envelope fails to exist over some intervals, be prepared for a few
error messages when generating the curve, as in Figure 15, where $\omega = 6$.

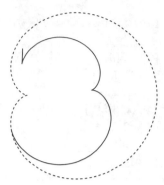

Figure 15. The numeral three as an envelope, $r = 1 + \cos\theta$, $\omega = 6$ (Exercise XIII.8)

9.   See Bivens [**11**, pp. 93–95] for a complete solution.

10.   See [**73**] for a complete solution.

Chapter XIV

1.   The analog of (VI.12) for distance dropped in time t for the ten pound ball is

$$s_1(t) = \frac{M_1}{\kappa_1} \ln\left(\text{sech}\left(t\frac{\sqrt{\kappa_1 g m_1}}{M_1} \right)\right),$$

where $g = 32$, $M_1 = 10/g$, $r_1 = 1/6$, $V_1 = 4/3\,\pi r_1^3$, $m_1 = M_1 - 2V_1$, $\kappa_1 = \pi k r_1^2$ for some constant k. Solve $s_1(20\text{ sec}) = -600$ feet for κ_1, getting $\kappa_1 \approx 0.009$, which means that $k \approx 0.026$.

Analogously define $s_2(t)$ for the distance dropped for the twenty pound ball, just as we did for $s_1(t)$. Note that $\kappa_2 = \pi k r_2^2$ where $k = 0.026$. Then Solve $s_2(17) = -600$ for r_2, getting 0.37 feet. That is, the second ball has radius 4.4 inches.

2.   Considering the graph in Figure II.5e, we may assume that the acceleration in the first few hundred miles of fall is almost a constant. Therefore we use the falling model $s(t) = \frac{1}{2}gt^2$ where $g = 32$ ft/s². So $s(3\text{ min}) \approx 98$ miles. The actual value will be less, since the ship is held up by air resistance.

3.   Let T be the time it takes Priscilla to traverse the launch tunnel. The equation for the exit velocity is $10gT = 7(5280)$, which gives $T = 115.5$ sec. Thus the tunnel has length $\frac{1}{2}(10g)T^2$, about 404 miles. (If we assume that the exit velocity is helped by the earth's rotation, we may reduce the tunnel's length somewhat.)

4. The only difference from the paths as given in Figures XII.8 amd XII.9 are the trajectory flight times.

5. This exercise is much like Exercise 3. Let $T = 5$ min be the time it takes Priscilla to traverse the 869 km launch tunnel length. Then $\frac{1}{2}\alpha T^2 = 869$ km, where α is a constant acceleration. Thus $\alpha \approx 19.1$ m/sec^2. Furthermore, the velocity of an object undergoing a constant acceleration of a for five minutes is αT. That is, Priscilla ripped through the crater fabric at a speed of 5.8 km/sec.

6. The numbers of days from day 0 to day 7.13.0.0.0 is $L = 7(20\,20\,360) + 13(20\,360) = 1101600$ days. The number of 365 day years in L days is $L/365$. The number of actual years in this length of time is two less than the number of 365 day years, namely, $L/365 - 2$. Let x be the length of an actual earth year. Then the number of earth years in L days times x should be L days. That is, $(L/365 - 2)x = L$. Solving for x gives the Mayan length of an earth year as 365.242 days.

7. Recall that in Chapter XIII, unit distance was chosen to be a light second, the distance that light travels in one second. Suppose for the moment that the central strand theatre is much larger, so that the length of the room is 2 light seconds and the distance between the two foci is 1 light second. Thus the eccentricty of the ellipse is $e = 1/2$ and the polar equation of the room is $r = 1.5/(2 + \cos\theta)$. The slowest rotation rate for which the origin is extraordinary is about 0.78 radians/sec, as illustrated in Figure 16. To generate this envelope, see the code in Exercise XIII.4.

Let $c = 186000$ mi/sec be the speed of light. In terms of light seconds, the polar equation of the room in the story is $r = (1.5/c)/(2 + \cos\theta)$ where r is in light seconds. Suppose that lake L has envelope Γ at angular rotation rate ω. Verify this conjecture: if L is scaled by the real number α, then the rotation rate which generates the envelope Γ scaled by α is the rotation rate ω/α. If this conjecture is true, then the minimum rotation rate which makes the origin an extraordinary point in the central strand is $0.78c$ radians/sec. This rate is very fast. Therefore the inscription on the beacon was misleading. Everyone in the room has a good view of the light show. However, the Grand Lunar may have superior reception, since the light from one focus reflecting off a smooth elliptical boundary passes through the other focus.

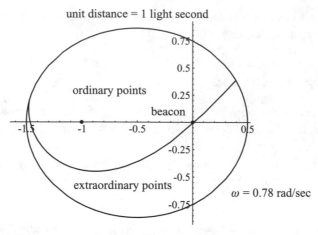

Figure 16. An envelope through the beacon, $r = 1.5/(2 + \cos\theta)$, (Exercise 7)

8. For (VII.5), take $k = 0$ since the moon exerts no gravitational acceleration on the cavorized submarine. Take $A = (R, 0, 0)$, $B = (-R, 0, 0)$ where $R = 1738$ km. Then $\mathbf{u} = \mathbf{i}$ and $\mathbf{w} = \mathbf{0}$. Thus $\lambda = -4\pi^2/Q^2$, $\mu = 0$, where $Q = 27.3$ days. By a generalization of (VII.8), the general solution to (VII.5) is $s = \alpha\cosh(2\pi t/Q + \beta)$ where α and β are real numbers. The initial conditions, $s(0) = R/2$ and $s'(0) = 0$ gives the solution

$$s(t) = \frac{R}{2}\cosh\left(\frac{2\pi t}{Q}\right).$$

Solve $R = s(t)$ for t to get $t \approx 5.72$ days. That is, it takes 5.7 days for Priscilla to exit the moon. Now evaluate $s'(5.7 \text{ days})$ for the exit velocity, namely 14.43 km/hr.

Chapter XV

1. 3.3×10^{10} km. This distance is well outside the solar system, as Pluto is 5.9×10^9 km from the sun.

2. Almost a month, about 29.6 days.

3. Kepler's third law says that the period of an orbiting body is $2\pi S^{\frac{3}{2}}/\sqrt{k}$, where S is the length of the semi-major axis of the ellipse. The path of the anvil falling from heaven's gate is a degenerate ellipse, whose semi-major axis has length $D/2$, and whose total falling time is half of the period of this degenerate elliptical orbit. By Kepler's law,

$$\frac{\pi(D/2)^{\frac{3}{2}}}{\sqrt{k}} = T,$$

the desired result.

10.* This is an open-ended problem. Experiment with various μ values.

References

1. Giorgio Abetti, *The History of Astronomy*, Henry Schuman, New York, 1952.

2. Milton Abramowitz and Irene A. Segun, editors, *Handbook of Mathematical Functions*, Dover, New York, 1968.

3. S. C. Althoen and J. F. Weidner, "Related rates and the speed of light," *The College Mathematics Journal*, 16 (1995) 186–189.

4. Peter Apian, *Cosmographia*, 1524.

5. Aristotle, *On the Heavens*, translated by W. K. C. Guthrie, Harvard University Press, Cambridge, 1958.

6. V. I. Arnol'd, *Huygens & Barrow, Newton & Hooke*, Birkhäuser, 1990.

7. Roger Bacon, *The Opus Majus*, translated by Robert Belle Burke, Russell & Russell, New York, 1962.

8. R. B. Banks, *Slicing Pizzas, Racing Turtles, and Further Adventures in Applied Mathematics*, Princeton University Press, Princeton, 1999.

9. William Beebee, "A round trip to Davy Jones's locker," *National Geographic Magazine*, LIX:6 (June 1931) 653–678.

10. E. T. Bell, *Men of Mathematics*, Simon & Schuster, New York, 1965.

11. Irl Bivens and A. J. Simoson, "Beholding a rotating beacon," *Mathematics Magazine*, 71:2 (1998) 83–104

12. Robert L. Borelli, Courtney S. Coleman, and Dana D. Hobson, "Poe's pendulum," *Mathematics Magazine*, 58:2 (1985) 78–83.

13. M. Born and E. Wolf, *Principles of Optics*, Pergamon Press, Oxford, 1965.

14. J. L. Brenner, "An elementary approach to $y'' = -y$," *American Mathematical Monthly*, 95:4 (1988) 344.

15. J. W. Bruce and P. J. Giblin, *Curves and Singularities*, Cambridge University Press, Cambridge, UK, 1984

16. Edgar Rice Burroughs, *At the Earth's Core*, Dover Publications, 2001.

17. R. Frank Busby, *Manned Submersibles*, Office of the Oceanographer of the Navy, 1976.

18. Lewis Carroll, *The Annotated Alice*, edited by Martin Gardner, W. W. Norton, 2000.

19. Chuang Tzu, *Chuang Tzu: Basic Writings*, translated by Burton Watson, Columbia University Press, Cambridge, 1980.

20. Arthur C. Clarke, *2001: a Space Odyssey*, The New American Library, 1968.

21. ———, *3001: the Final Odyssey*, Ballantine Books, New York, 1997.

22. Dante, *The Divine Comedy I: Hell*, translated by Dorothy Sayers, Penguin, 1949.

23. Leonardo da Vinci,*The Notebooks of Leonardo da Vinci*, edited by Edward MacCurdy, Konecky & Konecky, Old Saybrook, Connecticut, 2004.

24. Antoine de Saint-Exupéry, *The Little Prince*, translated by Katherine Woods, Harcourt, Brace & World, New York, 1943.

25. Ze-Li Dou and Susan G. Staples, "Maximizing the arclength in the cannonball problem," *The College Mathematics Journal*, 30:1 (1999) 44–45.

26. Arthur Conan Doyle, "The Stone of Boxman's Drift," in *Uncollected Stories: the Unknown Conan Doyle*, Doubleday, New York, 1982.

27. C. H. Edwards and D. E. Penney, *Calculus: Early Transcendentals Version*, Prentice Hall, Upper Saddle River, New Jersey, 2003.

28. Leonhard Euler, *Letters to a German Princess*, reprint of the 1833 edition, Arno Press, 1975.

29. Frank A. Farris, "Wheels on wheels on wheels—surprising symmetry," *Mathematics Magazine*, 69:3 (1996), 185–189.

30. Gary Feuerstein, Tower of Pisa painting, artist unknown, www.endex.com/gf/buildings.

31. R. P. Feynman, R. B. Leighton, and M. Sands, *The Feynman Lectures on Physics*, vol. ii, Addison-Wesley, 1964.

32. Camille Flammarion, "A hole through the earth," *The Strand Magazine*, 38 (1909) 349–355.

33. G. R. Fowles and G. L. Cassiday, *Analytical Mechanics*, Saunders, 1999.

34. M. Freeman and P. Palffy-Muhoray, "On mathematical and physical ladders," *American Journal of Physics*, 53:3 (1985) 276–277.

35. Galileo Galilei, *Dialogue Concerning the Two Chief World Systems—Ptolemaic & Copernican*, translated by Stillman Drake, University of California Press, 1967.

36. Chuck Groetsch, "From medieval scholasticism to cosmic exploration in three problems," *Primus*, 15:3 (2005) 215–225.

37. Stephen Hawking, *A Brief History of Time*, Bantam, New York, 1990.

38. LaDawn Haws and Terry Kiser, "Exploring the brachistochrone problem," *American Mathematical Monthly*, 102:4 (1995) 328–336.

39. Thomas Heath, *The Copernicus of Antiquity: Aristarchus of Samos*, McMillan, New York, 1920.

40. Hesiod, *The Theogeny*, in *Hesiod, the Homeric Hymns and Homerica*, translated by H. G. Evelyn-White, Harvard University Press, 1959.

41. William Heytesbury, "Rules for Solving Sophisms (1335) Part VI," translated by Ernest Moody, Document 5.1, in Marshall Clagett, *Science of Mechanics in the Middle Ages,* University of Wisconsin Press, Madison, 1959.

42. Keigo Iizuka, *Engineering Optics*, Springer-Verlag, 1985.

43. Nils P. Johnson, "The brachistochrone problem," *The College Mathematics Journal*, 35:3 (2004) 192–197.

44. Johannes Kepler, *Somnium, Sive Astronomia Lunaris*, translated by P. F. Kirkwood, in John Lear, *Kepler's Dream*, University of California Press, 1965.

45. N. Kollerstrom, "The hollow world of Edmond Halley," *Journal for the History of Astronomy*, 23:3 (1992) 185–192.

46. ———, "Newton's 1702 Lunar Theory," www.ucl.ac.uk/sts/kollrstm/hollow.htm, 2001.

47. Stanislaw Lem, *The Invincible*, Seabury Press, 1973.

48. Tung-Po Lin and R. C. Lyness, "Fast tunnels through the earth, problem 5614," *American Mathematical Monthly*, 76:6 (1969) 708–709.

49. D. A. Linwood, "The calculus and Gamow's theory of gravitation," *The College Mathematics Journal*, 31:4 (2000) 281–285.

50. Lyle N. Long and Howard Weiss, "The velocity dependence of aerodynamic drag," *American Mathematical Monthly*, 106:2 (1999) 127–135.

51. William Lowrie, *Fundamentals of Geophysics,* Cambridge University Press, 1997.

52. J. B. Marion and S. T. Thornton, *Classical Dynamics of Particles and Systems*, Saunders, 1995.

53. Brad Matsen, *Descent*, Pantheon Books, New York, 2005.

54. Peter M. Maurer, "A rose is a rose," *American Mathematical Monthly*, 94:7 (1987) 631–645.

55. A. A. Milne, *The World of Pooh*, E. P. Dutton, New York, 1957.

56. John Milton, *Paradise Lost: an authoritative text, backgrounds and sources, criticism*, edited by Scott Elledge, W. W. Norton, 1993.

57. NASA, *Space Educator's Handbook, (NASA Report Number S677)*, Johnson Space Center, Houston, 2006, www1.jsc.nasa.gov/er/seh/earlysf.html.

58. Isaac Newton, *Sir Isaac Newton's Mathematical Principles of Natural Philosophy and his System of the World*, translated by Andrew Motte, and revised by Florian Cajori, University of California Press, 1960.

59. Keith B. Oldham and Jerome Spanier, *The Fractional Calculus*, Academic Press, New York, 1974.

60. John Oprea, "Geometry and the Foucault pendulum," *American Mathematical Monthly*, 102:6 (1995) 515–522.

61. Ovid, *Metamorphoses,* translated by Samuel Garth, John Dryden, et al., in the Internet Classics Archive, classics.mit.edu/Ovid/metam.html, 2005.

62. Plato, *Euthyphro, Apology, Crito, Phaedo, Phaedrus,* translated by H. N. Fowler, Wm. Heinemann Ltd., 1960.

63. Plutarch, *Plutarch's Moralia,* vol. xii, translated by H. Cherniss and W. C. Helmbold, Harvard University Press, Cambridge, MA, 1957.

64. Edgar Allan Poe, "The pit and the pendulum," *The Collected Tales and Poems of Edgar Allan Poe,* Random House, New York, 1992.

65. Bruce Pourciau, "Reading the master: Newton and the birth of celestial mechanics," *American Mathematical Monthly,* 101:1 (1997) 1–19.

66. B. Ross, "A brief history and exposition of the fractional calculus," in *Lecture Notes in Mathematics,* vol. 457, Springer-Verlag, 1975.

67. M. A. Rothman, "Things that go faster than light," *Scientific American,* 203 (1960) 142–152.

68. Rudy Rucker, *The Hollow Earth,* William Morrow, New York, 1990.

69. R. Sahai (JPL) et al., Hubble Heritage Team (STScI/AURA), NASA, *Astronomy Picture of the Day, December 14, 2002,* apod.gsfc.nasa.gov/apod/ap041017.html/apod.gsfc.nasa.gov/apod/apo41017.html.

70. Paul Scholten and A. J. Simoson, "The falling ladder paradox," *The College Mathematics Journal,* 27:1 (1996) 49–54.

71. G. Shortley and D. Williams, *Elements of Physics,* Prentice-Hall, Englewood Cliffs, N.J., 1971.

72. G. F. Simmons, *Differential Equations,* McGraw-Hill, 1972.

73. A. J. Simoson, "An envelope for a spirograph," *The College Mathematics Journal,* 28:2 (1997) 134–139.

74. ———, "The trochoid as a tack in a bungee cord," *Mathematics Magazine,* 73:3 (2000) 171–184.

75. ———, "The gravity of Hades," *Mathematics Magazine,* 75:5 (2002) 335–350.

76. ———, "Falling down a hole through the earth," *Mathematics Magazine,* 77:3 (2004) 171–189.

77. ———, "Bringing Mr. H.G. Wells back from the moon," *Primus,* 14:4 (2004) 328–344.

78. ———, "Retrieving Mr. H.G. Wells from the ocean floor," *Primus,* 15:4 (2005) 363–373.

79. ———, "Sliding along a chord through a rotating earth," *American Mathematical Monthly,* 113:10 (2006) 922–928.

80. ———, "Playing ball in a space station," *The College Mathematics Journal,* 37:5 (2006) 334–343.

81. David Singmaster, *Bibliography on falling down a hole through the earth,* 87 Rodenhurst Road, London, SW4 8AF, England, singmast@lsbu.ac.uk.

82. William Smith, ed., *A Dictionary of Greek and Roman Antiquities*, John Murray, London, 1875.

83. Spiritweb.org, *Map of Agharta*, http://www.spirweb.org, 2001.

84. Frank D. Stacey, *Physics of the Earth*, John Wiley & Sons, New York, 1969.

85. G. Strang, *Calculus*, Wellesley Cambridge Press, Wellesley, Mass., 1991.

86. Ivor Thomas, editor, "From Aristarchus to Pappus," in *Selections Illustrating the History of Greek Mathematics*, vol. ii, Harvard University Press, Cambridge, Mass., 1957.

87. J. R. R. Tolkien, *The Two Towers, The Lord of the Rings*, vol. ii, Houghton Mifflin, 1994.

88. Alice Turner, *The History of Hell*, Harcourt Brace, 1993.

89. Henrik van Etten, *Recreation Mathématique*, 1624, as cited in Singmaster [**81**].

90. Jules Verne, *From the Earth to the Moon* and *Round the Moon*, in *The Omnibus*, J.B. Lippincott Company, Philadelphia, no date.

91. ———, *Journey to the Center of the Earth*, Tor Books, 1992. Frontispiece at http://JV.Gilead.org.il/vt/c_earth, 2001.

92. Voltaire, "Essays," in *The Works of Voltaire*, vol. xxxvii, Werner Co., Akron, OH, 1904.

93. H. G. Wells, "The Abyss," in H. M. Tomlinson, editor, *Great Sea Stories of all Nations*, Doubleday, 1937.

94. ———, *The First Men in the Moon*, the first edition, George Newnes, Ltd., London, 1901.

95. ———, *The First Men in the Moon*, in *Seven Famous Novels*, Garden City Publishing, New York, 1934.

96. E. Wigner, "The unreasonable effectiveness of mathematics in the natural sciences," *Communications on Pure and Applied Mathematics*, 13:1 (1960) 1–14.

97. Samuel H. Williamson, "What is the relative value?" *Economic History Services*, April 15, 2004, http:/www.eh.net/hmit/compare/.

98. Stephen Wirkus, et al., "How to pump a swing," *The College Mathematics Journal*, 29:4 (1998) 266–175.

99. M. Zeilik & E. v. P. Smith, *Introductory Astronomy and Astrophysics*, Saunders, 1987.

About the Author

Andrew J. Simoson was born in Minnesota, taught American Red Cross life-saving on the city beach of Escanaba on Lake Michigan for three summers, earned a BS in mathematics from Oral Roberts University in 1975, and a PhD in mathematics under Leonard Asimow at the University of Wyoming in 1979, working on extensions of separating theorems in functional analysis. Since then, he has been chairman of the mathematics department at King College in Bristol, Tennessee, and has authored over thirty papers in various mathematical journals, seven of them being joint research with undergraduates. The paper, "The Gravity of Hades," which is essentially Chapter II of this text, won the Chauvenet Prize for expository writing in 2007. He has twice been a Fulbright professor, at the University of Botswana, 1990–91, and at the University of Dar es Salaam in Tanzania, 1997–98. Having two sons, he was a long-time Cub Scout den leader, building model rockets, making marionettes and directing skits, leading camping and canoeing expeditions. He has also been a long-time youth soccer coach, and refereed many games. He and his wife regularly teach a college ballroom dance class each spring term. He is currently a member of the MAA and is an editorial board member for *Primus*, a journal on undergraduate mathematics teaching.

Index